HOW TO STUDY SCIENCE

FRED DREWES

Professor of Biology
Suffolk County Community College

 Wm. C. Brown Publishers

Book Team

Editor *Kevin Kane*
Developmental Editor *Megan Johnson*
Production Coordinator *Deborah Donner*

 Wm. C. Brown Publishers

President *G. Franklin Lewis*
Vice President, Publisher *George Wm. Bergquist*
Vice President, Operations and Production *Beverly Kolz*
National Sales Manager *Virginia S. Moffat*
Group Sales Manager *Vincent R. Di Blasi*
Vice President, Editor in Chief *Edward G. Jaffe*
Executive Editor *Earl McPeek*
Marketing Manager *Paul Ducham*
Advertising Manager *Amy Schmitz*
Managing Editor, Production *Colleen A. Yonda*
Manager of Visuals and Design *Faye M. Schilling*
Production Editorial Manager *Julie A. Kennedy*
Production Editorial Manager *Ann Fuerste*
Publishing Services Manager *Karen J. Slaght*

WCB Group

President and Chief Executive Officer *Mark C. Falb*
Chairman of the Board *Wm. C. Brown*

Cover and interior design by John R. Rokusek

Cover image © Manfred Kage/Peter Arnold, Inc.

Copyedited by Carol Danielson

Printed in the United States of America by Wm. C. Brown Publishers,
2460 Kerper Boulevard, Dubuque, IA 52001

10 9 8 7 6 5 4 3 2 1

D E D I C A T I O N

My wife, Sandy, has made this book possible. Sandy's dedication and sensitivity in the area of developmental education has helped me focus my attention on the needs of underprepared students. I would like to dedicate this book to Sandy for her help, encouragement, and patience during the preparation of this book.

CONTENTS

Preface, vii
Acknowledgments, viii

1 Study of Science, 1
Study of science, 1
Technology, 2
Your study of science, 2
Climate for learning, 3
Self-evaluation, 3
The rewards of studying, 4

2 Science Course—Instructors and Classes, 7
Science courses, 7
Lecture, 8
Types of instructors, 8
Course outline and requirements, 12
Laboratory, 12
Recitation, 12
Tutoring, 13
Study guides, 13
Science course survey, 13
Review, 14

3 Prerequisites to Study Science, 15
Prerequisites, 15
Review, 17

4 The First Day and 2 D's, 19
Front and center, 19
Classmates, 20
Instructor's expectations, 20
Discipline and direction, 21
Review, 21

5 This Study Guide—A Model, 23
Model for notebook, 25
Use of guide, 27
Review, 27

6 Study Skills Inventory, 29
Twenty-Seven Study Skills, 29

7 Time-Management and Study Sessions, 31
Time management, 33
Study sessions, 37
Review, 47

8 Class Notes, 49
Notes, 51
Listening, 51
Notebook, 53
Additional hints, 57
Review, 59

9 Use of Textbooks, 61
Textbook organization, 63
Survey, Question, Read, Recite, Review, and *Practice,* 65
Review, 69

10 Words and Symbols, 71
Words and symbols, 73
Review, 79

11 Figures and How to Use Them, 81
Types of figures, 83
Figure analysis, 89
Review, 91

12 Problem-solving, 93
Process of problem-solving, 95
Essay and math questions, 95
Math-based problems, 99
Review, 101

13   Types of Tests—Preparation, 103
        Types of tests, 103
        Preparation, 107
        Review, 111

14   Test-taking—Analyzing Results, 113
        Taking tests, 113
        Test-taking Skills and Hints, 117
        Analysis of test results, 121
        Review, 125

Appendices, 127

Bibliography, 135

PREFACE

To learn science, you must first *want* to learn it. A teacher can lead you to the subject matter but cannot make you learn!

Successful students have learned how to learn. They realize that they must do all of the following:

1. identify the objectives and standards of the course;
2. direct their academic efforts toward meeting the standards established by the instructor;
3. be able to predict what they must know;
4. use different skills in an organized and planned sequence of study;
5. generate appropriate questions about course content and confirm answers to these questions;
6. continually monitor their progress.

If you are uneasy about learning science, then this book will help you. However, it will help you only if you incorporate good study skills into your academic routine. These skills must become second nature to you; you shouldn't have to think about learning to learn. This guide focuses study skills on learning science, but obviously these skills can and should be used in other disciplines. How helpful the guide will be depends upon your attitude, academic background, and present level of study skills.

Chapters 1 through 4 talk a bit about the necessity of studying science, especially in college. It is important for you to become familiar with the nature and structure of science courses, because this knowledge will help you adapt to the challenge of these courses. This awareness will allow you to determine and control your academic behavior.

An inventory of study skills is listed in chapter 6. Compare your present study behavior to this skills list. This inventory will alert you to what you should do to learn science.

Chapters 5 and 7 through 14 discuss time management; study sessions; class notes; textbook use; words, symbols, and figures; problem-solving; preparing for and taking tests; and reviewing tests. All are important. You should read chapter 5, "This Study Guide—A Model" before studying the other chapters in the study skills section. Let your interests and needs determine which sequence of chapters you examine after that. Complete the study guide within the first two weeks of the semester, but refer to it later to support your study efforts.

The appendix contains answers to selected questions and tidbits of information related to study skills and learning. Note that the appendix lists sequences of study procedures. You might want to cut these out and use them as bookmarks or as a handy reference to a recommended study skill. I must emphasize, however, that the process of learning cannot be reduced to a cookbook list of ingredients. The lists provided here are simply guidelines to a pattern of study behavior.

I hope you will find this book helpful. I would appreciate any feedback you care to give. Thanks.

Fred Drewes
Suffolk Community College
533 College Road
Selden, NY 11784

A C K N O W L E D G M E N T S

I'd like to thank Ray Carlucci of Wm. C. Brown Publishers for his encouragement of me in getting this book started. Once started, various colleagues supported the effort in different ways. I thank Mario Caprio, Don Kisiel, and Paul Lauren. My family—Sandy, Kristen, and Brian—certainly deserve thanks for their support. Margaret Broderick, Terry Windus, and Pat Combattente deserve special thanks for their help during the typing of the book. Final thanks go to Francis M. Maxin (Community College, Allegheny College) and Ed Krol (Henry Ford Community College) for their helpful reviews and critiques.

1

Study of Science

Study of Science

Stars shine in the night sky. The moon waxes and wanes. The sun brightens our days. Clouds appear, then disappear. Rain might fall in a drizzle or a downpour. Water soaks into the ground or runs off into rivers. Clay, silt, sand, and gravel are formed and carried about. Plants absorb groundwater and sunlight. They grow, producing roots, stems, leaves, and flowers. Animals, including human beings, scurry about, claiming territory and struggling to survive. Each living organism produces its own kind after eggs and sperm of its species unite.

Human beings observe these and other things in our universe. Early peoples knew the ways of the local land, water, plants, and animals. They learned to domesticate plants and animals, to direct the flow of water, and to use the heat of fire to forge new tools. As the human species evolved and our numbers expanded, we developed areas of knowledge to explain the world and the universe to satisfy our hunger for knowledge. *Science is the discipline that studies the quantitative as well as the qualitative nature of the world.* Scientists attempt to answer *what, when, why, where,* and *how* questions; they try to understand and describe the order of the universe.

Why do stars shine? What are they made of? Why does the moon seem to change its shape and position throughout the month? Why does rain fall? What is water? Light? A flower? An animal? A human being? Where did they all come from? How does a fertilized egg program itself to develop into a tree, a toad, or a person? Why do things work the way they do? One question leads to another; that is the nature of science.

To seek answers to questions, the scientific method blends creative thinking, critical thinking, and problem-solving techniques. Scientists design *experiments to test hypothetical answers* to questions that have been asked about their observations. They develop tools with which to perform these experiments. The experimenters establish *procedures* to gather quantitative and qualitative *data*. This data is *analyzed,* and a *conclusion* is drawn that supports, modifies, or refutes the initial answer. After repeated experiments lead to the same conclusions, *theories* are formulated. These theories can then be used to explain past observations and to predict answers to new questions.

Scientists and students of science find science fascinating, exciting, and fulfilling. When they discover something new or arrive at a new understanding of a befuddling problem, they might shout out "Beautiful! Great!" or "Hot damn!" They might even fly some high fives.

Modern science and its quantification of the characteristics of the "stuff" of the universe started about 500 years ago. In the past 100 years, there has been an explosion of scientific inquiry. More recently, computers—developed from discoveries made by scientists—are being used to expand our investigation into the details of the universe even faster and farther. The inquiry is taking us from the tiny, infinitely detailed code of life's all-important hereditary molecule (DNA), to the vastness of space.

Technology

Technicians of all varieties use scientific information to make all manner of tools, machines, buildings—all of the things we call modern technology. We have progressed from the Stone Age to the computer age. Our constant hope is that all of this activity will help us explain the nature of the universe and allow us to live better, fuller, longer lives. Technology is not only an important part of our day-to-day life, but it also helps scientists extend their knowledge and understanding. This application of science to create technology obviously makes us dependent on science and on modern technology. We have become dependent on modern machines and medicines. And what would we do without the plentiful supply of energy provided by modern technology? Don't we expect it all to continue?

Scientific knowledge is neutral; it is neither good nor bad. The answers scientists seek are either right or wrong. How we use the science can have desirable or undesirable results. Bacteria and viruses can be grown to produce antibiotics, antibodies, or vaccines. They also can be grown to create biological weapons. Chemists create new and useful molecules, but some molecules turn out to be mutagenic, carcinogenic, or teratogenic. Physicists explore the mechanisms of matter and motion, and engineers put this information to use in farm and war machinery.

Your Study of Science

Because people are so dependent on science and technology, we all should

1. have some basic knowledge of the concepts and content of science;
2. be willing to learn science;
3. use scientific information and the scientific method to help us make decisions about the best way to conduct our lives.

The welfare of any modern, technological country depends on an informed and active citizenry. Scientific literacy is an important part of good citizenship. National issues must be resolved and public policy developed. Knowledge and communication of science are vital parts of many national issues. Abortion, fetal tissue research, and the future use of genetic code information are all issues that are vigorously debated. What new laws should be created? National energy use (the sum total of all people's direct and indirect energy use), the quantity and nature of air, water, and land pollutants, and our desire for greater and greater material wealth are set to converge and clash. How can we conduct our lives to resolve or manage these problems?

When we are young, we ask all sorts of questions: *why, what, when, how?* At a certain age, many people learn to stop asking questions, to stop seeking answers, to seemingly stop wanting to know. If this applies to your desire to learn science, you must once again become a kid at heart. Be ready to ask questions and seek answers. Try to drink deeply from the well of science.

As a student, you must fulfill certain requirements to earn your degree. You will invest effort, time, and money to get it. Some students possess the basic skills and have the motivation necessary to do well in the required science courses. Other students have not developed or have forgotten the study skills they need to succeed. Still other students lack the self-confidence to study science; they have "science anxiety."

If you feel you need help in learning the content of a science course, this book is designed for you. The recommendations are designed to be practical, and valuable. With regular *practice* and *application* of these study skills, you should succeed! These recommendations are gleaned from personal experience, observations of and interaction with successful and unsuccessful college students, and a review of other study skill guides.

Successful students work hard, concentrate on the task at hand, have an organized system of study, and have established a strong foundation. You can learn science if you take the time to learn how to learn. You will be able to do a good job at it if you give yourself the chance.

At this point, you might be what is called a concrete thinker. In an anatomy course, you would be able to learn the names of muscle groups with few problems. However, if the physiology (chemical nature) of muscle contraction were discussed, you would have difficulty learning the material. Physiological information and concepts are more abstract than rote memorization; the ability to think abstractly and to make interconnections is required. If, as a concrete thinker, you were asked to relate force, resistance, and effort to a math-based problem on muscle strength, you would experience difficulties in solving the problem. Again, the application of the principles of physics is abstract. Your level of thinking should be pretty close to the level of the course material presented. If it is not, then you will experience difficulty in learning.

Climate for Learning

Your ability or inability to succeed in college science courses is related to many different factors. Rate yourself on a scale of 1 (low) to 10 (high) for each factor listed below. Record your scores in the spaces provided.

Self-Evaluation

_____ **1.** General attitude toward learning.
_____ **2.** Attitude to learning science specifically.
_____ **3.** General environment for learning.
_____ **4.** Encouragement of family and peers.
_____ **5.** Motivation to spend the time and exert the effort necessary to learn science.
_____ **6.** Amount of study time available.
_____ **7.** Quality of study skills.
_____ **8.** Observational and problem-solving skills.
_____ **9.** Competency in reading, writing, and mathematics.
_____ **10.** Self-image and self-confidence levels.

_____ **11.** Comfort in learning science.
_____ **12.** Ability to concentrate on a topic.
_____ **13.** Self-discipline.
_____ **14.** Ability to direct your own study or work.

If you have rated yourself from 1 to 6 on any of the above items, then these areas need attention. How will you go about improving your attitudes and skills in these areas?

Remember, it is your responsibility to learn.

Figure 1.1   How does this figure relate to your attitude about learning?

The Rewards of Studying

The basic reason people do anything in life is for the rewards they gain from their actions. Let's face it; the two rewards for studying and learning the content of an introductory science course are for passing *grades and satisfaction.*

Several factors contribute to your desire and motivation to learn. A high grade, a good class standing, a scholarship, recognition, social approval, satisfaction of a natural curiosity, a need to feel competent all blend together in a complex way to be your reward for completing a science course. If you are highly motivated, then studying and learning will occur and rewards will be gained. If you are poorly motivated, then the amount of studying and learning will be scanty and rewards will not materialize. As a matter of fact, frustration will replace satisfaction. You should realize that some instructors, by the power of their personalities, might help motivate you or nudge you on to success, but *it is up to you to move yourself.*

Some people learn vast amounts of information about their hobby. For example, children learn fantastic amounts of information from baseball cards (a form of flash card). To them, owning the cards and learning all about the players is satisfying. As the youngsters mature, some may even study the strategy and concepts of game plans. They will then have progressed from concrete to abstract thinking, and they will be learning for the simple satisfaction of it. People also collect baseball cards as an investment, and the only statistics that are important to them are the name, date, and dollar value of the card. The value introduces a tangible reward somewhat similar to grades. Whatever the case, learning by the hobbyist is self-directed and self-disciplined. If this motivation could be bottled and sold, it would make millions.

You probably do not view the study of a college science course as a hobby. College science courses are requirements! Because courses are requirements, some of the joys a hobbyist experiences while learning might be absent. If you focus on keeping a positive attitude and the eventual reward of gaining knowledge rather than on a negative ("I can't") attitude and the feeling that you are wasting time and effort, then you might well receive better grades and a high level of satisfaction. Cultivate an interest in learning, delve into the content. Introductory science courses are challenging, and it is *gratifying* to do well in them. *The study skills recommended in this book will help you achieve these rewards.*

2

Science Course— Instructors and Classes

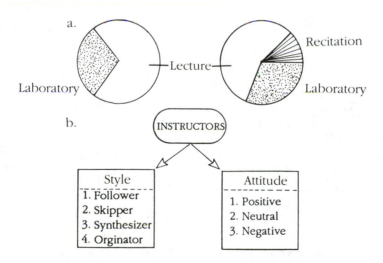

Figure 2.1 Science classes (*a*) and instructors (*b*).
Note (*a*) is an example of a pie graph.

Science Courses

Introductory science courses are made up of two or three components (figure 2.1). Science courses for *non-science students* have two or three lectures and one laboratory each week. Science majors' courses in chemistry and physics generally have a recitation or problem-solving class as well as lectures and laboratories. Astronomy, biology, botany, geology, and zoology generally have lectures and labs but no recitations. The lecture class size varies from around twenty to several hundred students. Laboratory classes also vary in size but generally have one instructor for fifteen to twenty-five students.

Lecture

The bulk of the course content is presented during the lecture by instructors. Most instructors write ideas or concepts on the board. *Remember:* The main body of the lecture discusses the relationships between these recorded points. *Notes* must be taken on the *discussion* as well as on what is recorded on the board. Instructors use a variety of visual aids. Because many book publishers provide transparencies as teaching aids, transparencies are probably the most frequently used visual aid. The figures on these transparencies are often the same as in the textbook. Be sure to record the figure number in your notes and *take notes on the information described in the transparency.* This is a challenging skill to learn.

Types of Instructors

Your instructor's style influences the way you listen, take notes, and study. An instructor's style of teaching can be categorized into one of four groups (figure 2.1). The four types of instructors are the follower, the skipper, the synthesizer, and the originator.

1. Follower
   a. Follows the content of a textbook chapter by chapter, page by page, example by example.
   b. Bring your book to class so you can check the sections of material covered. You should take notes on any additions or to clarify text material.
   c. You will find it easy to preview material before attending lectures. Your textbook with the checked material and notes are the guide to the material to be studied and learned.
2. Skipper
   a. Follows the content of a textbook but skips around from one part of the book to another.
   b. The skipper's content might or might not be easy to trace in the textbook. Note-taking is important to keep track of exactly what was covered and in what sequence.
   c. You will find it easy to preview material before attending lecture, because assigned readings are probably given. For study purposes, your lecture notes are as important as the text.
3. Synthesizer
   a. This lecture style draws on many different parts of the textbook and on outside resources. All is synthesized into the lecturer's view of the introductory science course. This view might introduce the same material as the text but in a completely different way.
   b. Lecture notes are very important because the topics cannot be quickly found in the text. Study of the notes and of similar information in the text helps solidify learning. The index of the textbook should be used to find the information corresponding to the lecture notes.
   c. This type of presentation will be difficult to preview unless the instructor indicates what areas will be covered in the lectures to come.

Box 2.1

This is an Example of an Outline

Organization of lectures

Instructors organize their lecture content in a number of ways. It will help you to listen and learn if you recognize the patterns usually used to organize the information.
   Types of organization

A. Chronological (in time)
   1. Sequence in which subject was seen or done
      Example: Study of the atom
   2. Sequence in which the subject should be seen or done
      Example: A laboratory experiment or
      the solution to a problem
   3. From cause to effect
      Example: Radiation causing
      mutations or cancer
B. Spacial
   1. What is next to what
      Example: Different strata of rock in
      the Grand Canyon
   2. What is connected to what
      Example: Description of the digestive
      system
C. From general to specific
   1. Theoretical to practical
      Example: $F = md$, Work of pulley
      systems
   2. General topic to examples
      Example: Function and nature of
      enzymes or catalysts to discussion
      of pepsin or platinum
D. From least to most (or most to least)
   1. Small to large
      Example: Atom to biosphere
   2. Weak to strong
      Example: Chemical bonding
   3. Simple to complex
      Example: Discussion of tissues or the
      structure of atoms
   4. Least controversial to more controversial
      Example: Discussion of origin of the
      universe

Adapted from *Study Smarts* by Judi Kesselman-Turkel and Franklynn Peterson (Contemporary Books, Inc. 1981).

Box 2.2

> **What is Important to Study?**
>
> *Here is a list to guide you as you pick out the important things to study.*
>
> 1. Anything mentioned in lecture or lab has a very high priority.
> 2. The more time that is allocated to a topic, the more time you should spend studying it.
> 3. If something is mentioned in lecture, lab, and the textbook, then this is very important material to learn.
> 4. You should be able to define scientific terms or symbols used in a lecture or lab and be able to use them in a conversation.
> 5. Learn the content of figures used to illustrate concepts, principles, processes, and facts.
> 6. Hints and gestures from instructors highlight important information.
> 7. Rhetorical questions your instructor asks are clues as to what you should learn.
> 8. Specific assignments indicate the material that is important.
> 9. Problems solved in lecture, lab, or recitation are samples of what you are expected to be able to solve. Practice solving similar problems.

4. Originator
   a. This type of instructor presents information from a variety of sources, much of it originating from recent publications. Collections of readings might be important, rather than a single text. Independent study, research, and a strong academic background are expected.
   b. Note-taking is important. Self-directed study of references is also important.
   c. If a textbook is recommended, you will probably use it as a reference, or you might wind up not using it at all.

All of your instructors also have a certain attitude about their jobs as teachers and about your job as a student. Some instructors display a very positive attitude toward teaching, an enthusiastic interest in the subject matter, and a sincere interest in whether you learn the subject. Other instructors are purely dispensers of information and seem to have a very neutral attitude toward job, subject matter, and you. Still other instructors are negative individuals; they are dissatisfied with their jobs and would rather be doing something else.

You, as a student, have a certain learning style. You also have a certain set of attitudes that influence your learning. Your own style of learning might mesh well with your instructor's style of teaching, but if it doesn't, you will have to take steps to cope with the situation. It is your job to learn the material; the educational system at college leaves it up to you to comprehend the subject matter. Some instructors make the material seem relatively easy and interesting; other instructors are sources of great frustration.

Box 2.3

How do self-satisfaction and grades relate to each other? Where would you place yourself on this graph?

Note that if you answered this question, you had to

1. Read the labels on each axis.
2. Choose a level of satisfaction.
3. Choose a grade.
4. Locate the level of satisfaction on the vertical scale; locate the grade you desire on the horizontal scale. Then draw an imaginary line from the satisfaction axis and up from the grade axis. Where these lines meet is where you placed yourself on the graph.

     If you have done this task successfully, you can use a graph. If you can't do this, then see a tutor about how to use and construct graphs.

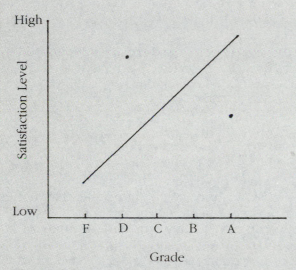

Figure 2.2   Relationships between the degree of satisfaction and earned grade. Note this is an example of a line graph but it is more qualitative than quantitative.

Generally, the degree of self-satisfaction a person feels is directly proportional to the grade he or she earns. The higher the grade, the greater your feeling of self-satisfaction. However, exceptions do exist. A student who never did well in high school might be very pleased to earn a "D" or "C" in a science course that is known to be difficult. On the other hand, an academically strong student might reap little satisfaction from earning an "A" in a class that is difficult.

Course Outlines and Requirements

Syllabus

Most instructors hand out a course outline on the first day of class. As the instructor reviews the outline, *have a pencil or pen in hand,* follow the review and *take notes.* The outline generally includes the following:

1. Course objectives
2. Title of required textbook and manual
3. Statement of teaching approach
4. Course outline
5. Reading assignments
6. Grading method and value of tests, assignments, and other work
7. Attendance policy
8. Makeup procedures
9. Tutoring center information
10. Policy on course withdrawal

Your final grade will be based solely on an average of test scores and graded assignments. The amount of effort and study time, class participation, and extra projects are seldom part of the final grade.

You will find that the instructors expect *you* to find out what *you* missed if *you* are absent. Your absence does not excuse you from having to learn the material. Exchange telephone numbers with a few classmates so you can call them if you must miss class.

Laboratory

Laboratories are meant to provide a direct, physical experience with the process of science. The information in the lab might or might not parallel that in the lecture. The laboratory instructors might or might not be the same as the lecturer. Four-year colleges and universities frequently hire graduate students to teach laboratories. Some laboratory instructors might just sit at the desk and answer questions, assuming that you are doing the work and understanding the material. Other instructors move about to check on how well you are doing the exercises. It is up to you to take the initiative to become involved in the lab work. Don't be afraid to call upon instructors for help. It is their job to answer your questions. You have paid for their assistance. However, it is a bit embarrassing to ask questions if you have not properly prepared for the class. You will use laboratory manuals to guide your lab study. Don't sit back and watch your lab partners do all the work. If you do, you will find they will learn the material more easily than you will.

Requirements for laboratory reports vary. Formal laboratory reports are generally required of students majoring in science. Nonscience courses might have a few such reports assigned, but more than likely they will require that the questions in the laboratory manual be answered. Be sure to clarify your responsibilities.

Recitation

A recitation class is devoted to problem-solving. You will use this time to review the solutions to problems assigned to you. This type of class reinforces the content of the lecture and laboratory classes.

Tutoring

Most colleges have organized tutoring services to assist you. The tutors are usually graduate or undergraduate students qualified to help guide your study. As with instructors, their level of skill will vary. If you have problems grasping the course content, you should definitely use the tutoring service. In addition, don't be afraid to *seek out the instructor's help.* Make an appointment before or after class. Generate questions before you seek their help. The questions will be a valuable way to start a tutoring session.

Study Guides

Some publishers provide computerized study guides for the textbook. These might be on file in the learning center of your college. Study groups find this type of guided learning to be very helpful to the individuals in the group. If your instructor is a follower or a skipper, then these guides will be a valuable study aid. Ask your instructor whether they are available.

Printed study guides, available in the bookstore, are also available as companions to your text. *They are helpful if your instructor is a follower or a skipper.* Ask your instructor for a recommendation.

Science Course Survey

1. How many science lecture hours do you have each week?
2. How long is your lab?
3. What is the teaching style and attitude of your instructors?
4. What is your academic attitude and learning style?
5. Who wrote the textbook and laboratory manual you use in the course?
6. Is there a limit to the number of absences you may have in lecture? In lab?
7. How many tests will be given?
8. What types of tests are given?
9. What kinds of laboratory assignments are given?
10. What is the proper format of a formal laboratory report?
11. Are any research papers required? When are they due?
12. Where and when are tutoring services available?
13. What are the office hours of your instructors?
14. Does the course outline include detailed reading assignments?
15. What kinds of visual aids does the instructor tend to use?
16. What kinds of notes does the instructor record on the board?
17. When problems are solved, are detailed explanations given?
18. How many hours do you need to spend each week learning the subject?
19. Where do you sit in the classroom?
20. What does your body language say about your attitude and desire to learn science?
21. How is your final grade determined?

You will be exposed to many different academic stimuli in the science course (figure 2.3). You are expected to learn the material and then demonstrate your comprehension of it. Your responses to different forms of evaluation result in a grade. You will develop certain feelings about the overall course of study. In the end, you will judge your overall level of satisfaction.

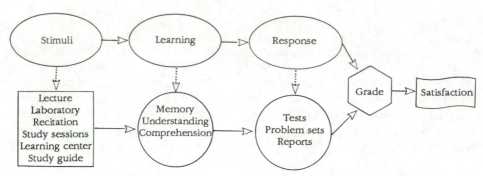

Figure 2.3   The stimuli in the course will have to be processed and used to lead to satisfaction.

Review

1.   Science courses have lecture and laboratory classes each week. Some sciences have a recitation class each week.
2.   Instructors might be followers, skippers, synthesizers, or originators. How you take notes and how you study are influenced by the instructor's style.
3.   Instructors have attitudes that influence their effectiveness as teachers. It is your job to comprehend course content regardless of the instructor's teaching style and attitude.
4.   Use instructors and tutors to help you answer questions you can't resolve. Instructors and tutors are paid to help you.
5.   The grades in the science course are based on your performance on various tests and graded assignments.

3

Prerequisites to Study Science

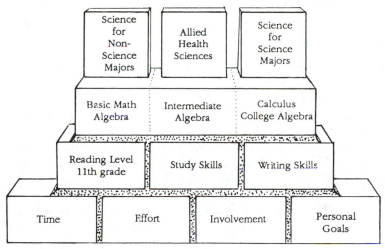

Figure 3.1   Foundation needed to study science.

Prerequisites

You will need to meet certain prerequisites to successfully study college science courses. Before or as you begin to study a science course, you should evaluate yourself to see if you have the foundation to succeed in the course (figure 3.1). If your self-analysis indicates that you have a good foundation, then you can feel confident about taking the course. On the other hand, if your analysis indicates various deficiencies, then you must either strengthen your academic foundation or cope with the difficulties and frustrations you will experience.

Box 3.1

Support of peers and family

You will find it helpful if the people around you support your efforts to learn. As you become involved in learning, you will want to share the information you have worked so hard to learn. Talking about what you learn reinforces your learning.

Unfortunately, friends and family might not want to listen to you. They might have a negative attitude toward learning. They might also feel threatened by your mental activity or the prospect of your having a college education. In fact, peers and family might make learning difficult. That will be frustrating! Thus, learning can be risky because personal relationships might change or even become endangered. Learning is also risky because you are opening yourself up to evaluation. You might perceive the evaluation as a potential chance to fail. This self-exposure is risky to people with low self-esteem.

Study group

Find a few classmates who are also learning. Get to know them and exchange phone numbers. Ask if they would like to form an academic support group. The group should meet weekly to study. This group work is refreshing and rewarding. If you are outgoing, then this will be easy to do. However, if you're somewhat shy, it will take gumption on your part to come out of your shell.

How well you do in a science course depends upon several factors:

1. *Personal standards.* Set a specific level of achievement and then do the work to reach it. Instructors enjoy having academically aggressive students in their class.
2. *Involvement.* Be an active learner, ask questions, participate in labs. It is important that *you want to learn and to show it.*
3. *Effort.* Learning science takes work. Break down the big tasks into smaller, manageable tasks, thus spreading out the work. Use different study skills to study the same material.
4. *Time.* Experience has shown that six to ten hours per week per class are needed to achieve a grade of *C* to *A*. How much time you will need depends on your study skills and academic background. How much time you will actually spend studying depends upon your standards, commitment, job requirements, social life, and health. If you don't have enough time, then don't bother taking the course.
5. *Study skills.* Your ability to listen, observe, question, take notes, concentrate, use a textbook, work in lab, organize time and tasks, solve problems, and take tests influence how easy or difficult your study will be. These skills are separate but related to your ability to read, write, and solve math problems. Science is a quantitative and detailed analysis of the universe. Symbols and unique vocabulary are used to convey this information. These must be learned before communication and comprehension can begin.

6. *Academic skills.* If you have an eleventh-grade reading level or higher, can write competently, and understand basic mathematics, then you will be able to handle an introductory science course for non-science majors. Biology, chemistry, and physics majors require competency in college algebra or calculus. You can determine the level of your reading, writing, and mathematical abilities by taking various competency tests. See college counselors to arrange to do this if it is not already part of a testing program following your acceptance to a college. Extra time and effort might compensate for a poor foundation, but it will be tough. Strengthen your basic skills; build a stronger foundation before you pursue your college studies.

A college degree is worth achieving. The study of science is one of the basic requirements for such a degree. You can study science successfully and earn your degree if you have good study skills. There is no reason why you can't develop good study skills.

Review

To successfully study science, you must

1. establish a personal standard of achievement;
2. become involved in the study of the science;
3. work at learning;
4. have and spend time to learn;
5. possess and use good study skills;
6. have competency in reading and writing equivalent to eleventh- or twelfth-grade skills;
7. have competency in mathematics compatible with the science being studied.

4

The First Day
and Two D's

THE FAR SIDE   By GARY LARSON

"Well, I've got your final grades ready, although
I'm afraid not everyone here will be moving up."

Figure 4.1   Where to sit?
What to do? (The Far Side.
Copyright 1991 Universal Press
Syndicate. Reprinted with
permission.)

Front and Center

Decide to sit toward the front and center of the classroom. Claim that territory.
It will be yours for the rest of the semester. There are a number of advantages
to this. You will

1.   find it easier to concentrate on listening to the lecturer;
2.   be able to see the board and projection screen better;
3.   not be distracted as much by casual conversations;

4. not be distracted as much by the beauties and the beasts in the class;
5. be less likely to escape into your own daydreams;
6. get handouts and tests sooner.

By sitting in front, you are acting on your decision to be an active, attentive student in the class and to become involved. Sitting near the front might be difficult to do, but *do it!*

For the first and all lecture classes, you should bring pencils or pens, a notebook (preferably a looseleaf), and your textbook. *Get to class a few minutes early, open your notebook, and get ready to start before your instructor begins.* As the semester continues, you will be able to use these few moments to review your notes. For lab, bring pen, pencil, notebook, textbook, and lab manual. As the term progresses, you might find that you do not need the text, but it is always a handy reference to have.

If on the first day your instructor presents an overview of the course, *take notes.* Don't just sit back and listen, thinking this summary is unimportant or that you will remember it. You can gain some important clues about the entire course on the first day. In addition, your instructor will start to display his or her style and standards. You can begin to practice your *listening and note-taking skills.* You can begin to gauge what kind of challenge the science course will be.

Classmates

Start to get acquainted with your classmates. Open a conversation with them. Begin to determine who might have similar goals and attitudes toward study. Try to find a group of people interested in forming a study group. Some students seem to be intimidated by the behavior or apparent knowledge of other students. Try not to be intimidated. Do your own thing. Study and learn. Take control of yourself and your actions. Remember, though—*group study* is another important study skill.

Instructor's Expectations

Instructors expect self-disciplined and self-directed study. Reading assignments are given. It is up to you to decide when and how thoroughly you study the assigned and unassigned material. Often instructors will say, "Answer the questions at the end of the chapter or the end of the lab exercise." That is the last you might hear about the assignment. You are just expected to take the *initiative* to answer the questions. The next time you see the questions, they might be on a test.

Instructors assume you will

1. know how to study;
2. have an appropriate academic background;
3. study until you learn the course content;
4. seek help if you need it;
5. know how to determine what to study;
6. be self-motivated;
7. be self-disciplined;
8. be self-directed.

Discipline and Direction

Students indicate that two difficult adjustments to college life involve *self-discipline and self-direction.*

Academic self-discipline is the ability to establish and follow a set of rules and regulations to guide one's academic behavior. Although each college has its own set of behavior standards, none apply to the individual's time management and application of study skills. Most students have no problem behaving in a socially acceptable fashion. The big problem is how to behave in an *academically* acceptable fashion. As a student, you must develop rules of study and stick to them. This requires self-discipline. No parents, no teachers to remind you. You have to have the motivation to get yourself to do the work. You have to set the alarm clock and get up. You have to go to class and actively listen and learn. You have to arrange to make up missed work. You have to set the time to study efficiently and effectively.

Self-direction is the ability to decide when, what, where, and how much you should do. It is a behavior pattern established by yourself for yourself. It is a pattern that directs you to establish certain objectives and allows you to accomplish and achieve the objectives. If you have trouble directing yourself, find out how others direct themselves. Ask how they give themselves direction to study. If you join or form a study group, then the group will tend to give you direction. Don't be afraid to seek help and encouragement from your peers.

To sum up:

- Establish academic objectives.
- Develop a sense of satisfaction when you have achieved these objectives.
- Realize that your study will generally have long-term rather than short-term benefits.
- Combine social and academic activities. (Have study dates.)

Review

1. Sitting near the front and center of the classroom helps you be attentive.
2. Get to know some of your classmates. Form a study group.
3. It is your responsibility to learn the material presented in class; you must take the initiative to become an active learner.
4. Develop self-discipline and self-direction.
5. "The ball is in your court."

QUESTIONS	ADDED CONTENT
• What is motivation?	Motivation can be defined as the desire to be competent. It is the inner drive in people that gets them to accomplish a task. Motivation is the quality that gives a person self-discipline and self-direction.
• What will get you to use this guide?	People are generally motivated to do the things they consider valuable to themselves. A person who considers nothing to be of value is in fact poorly motivated.

• What are the different kinds of problems college students experience?

Common problems of college students:

1. Conflicts of values with peers and friends.
2. Not knowing what to expect and what is expected of them.
3. Adjusting to dormitory life.
4. Adjusting to responsibility for one's own actions.
5. Adjusting to an independent life-style while still living at home.
6. Making own decisions.
7. Missing friends and family.
8. Money.
9. Work.
10. Time—too much or too little.

5

This Study Guide—
A Model

Objective

1. To learn how to use this book.
2. To understand that the format of this chapter is a model for your science notebook.

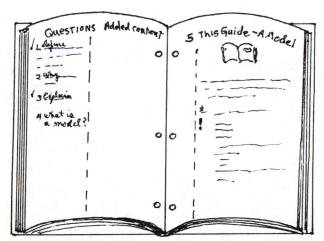

Figure 5.1 Organization of the study guide and a model for your notes.

Q U E S T I O N S	A D D E D C O N T E N T

- Explain how this book is organized.

- Why are the pages of the remaining chapters organized the way they are?

- How does your present note-taking style differ from the format of this chapter or the style suggested in the note-taking chapter?

- Why should you bother to think up questions about the content of the course you are studying?

- Does it make sense to think up your own questions? Why or why not?

- Which three chapters of this book do you think will be most helpful to you?

- What are the differences between figure 5.1 and 8.1?

- Does figure 5.1 accurately represent the first right-hand page of this chapter?

Model for Notebook

The format of this chapter and of the remaining chapters serves as a *model* for how you should organize your notebook. A notebook should be a tool you use for studying. Often notes are a jumble of unconnected words, phrases, symbols, and misinformation. By the time you get to the note-taking chapter (page 49), you will probably be sensitive to the fact that how you organize your notes, how much room you leave for corrections and additions, and what questions you record are all necessary to turn a jumble into a useful study tool.

Chapter 8 discusses lecture notes more thoroughly, and figure 8.1 diagrams the recommended notebook format. Take a moment to compare the format of the chapters in this book to figures 5.1 and 8.1. Note that the remaining chapters begin with a *title,* a *list of objectives,* and a *figure.* These elements summarize what the chapter is about. Below each figure appears the *content* of the chapter. This contains the pearls of wisdom and the rationale that will help you develop the study skills you need to succeed in science courses. A chapter *review* highlights the important parts of the chapter.

The *left-hand page* is divided into *questions* and *added content* columns. Questions are asked to focus attention on the content of the chapter and to test your knowledge. It is important—no, vital—that you develop the ability to generate questions about the science material you are studying. This technique will help you identify what you know and what you don't know. Asking and answering questions will help you monitor and gauge your learning progress. Remember, when you were a child you probably asked question after question to learn about the things around you.

The *added content* column might have either additional information or be blank. If there is blank space, you can use the space to write notes or comments to yourself, or to answer the questions that appear in the questions column.

DENNIS THE MENACE

"I GOT A LOT OF ANSWERS, BUT NO ONE EVER ASKS THE RIGHT QUESTIONS."

Figure 5.2   An illustration to make a point . . . asking the right questions is important.
("DENNIS THE MENACE"® used by permission of Hank Ketcham and © by North America Syndicate.)

Q U E S T I O N S

- In this chapter, three methods are used to explain the best way to use this book. They are the written outline, a flow chart, and an information map.

- Which is easiest to read?

- Which is easiest to quickly review?

- Do you make up flow charts or information maps to

  1. organize sequences of information given in lecture?
  2. reorganize minor information as it relates to major concepts?
  3. establish what will be done during lab experiments?

A D D E D   C O N T E N T

Sometimes it helps to draw a figure about a discussion in lecture, lab, or text. Here is an example of a flow chart and an information map on how to use this book.

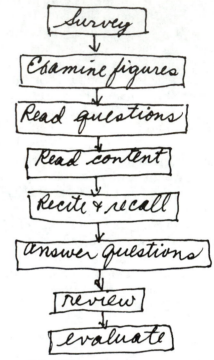

Figure 5.3   Flow chart. How to use this book.

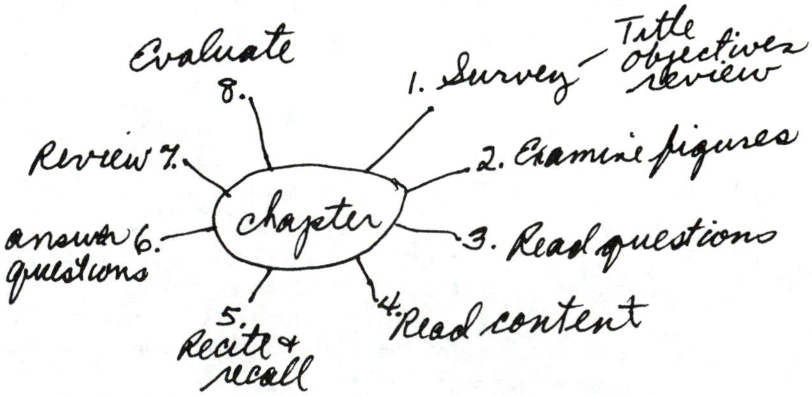

Figure 5.4   Information map. How to use this book.

Use of Guide

To use this guide, you should

1. Survey the chapter by reading the title, objectives, and the chapter review.
2. With pencil in hand, examine the contents of the figure. Identify the important parts and their relationship to each other.
3. Read the questions in the questions column. Formulate questions of your own. Use how, what, when, where, why, explain, and so forth to generate questions.
4. Read the content of the chapter.
5. Recite or recall the information in the chapter.
6. Answer the questions asked in the questions column.
7. Review the objectives, the chapter review, and the figure.
8. Evaluate what has been presented. Think about it. Adopt and practice these study skills as you study for your science course or any other course.

Reading what is written in this book will not help you unless you use and refine these study skills. If you are thinking, "Yes, I'll do it if I have the time. I really will try if I can squeeze it in after work," put the book down, get your money back, and go and do something else. *Say to yourself,* "I will develop these study skills. I will put them into practice."

Review

1. The format of this book is a model of how you should organize your lecture notes.
2. Each chapter has a title, a set of objectives, a figure, chapter content, a review, a questions column, and an added content column.
3. Each chapter should be surveyed, questions asked, content read, questions answered by reciting or writing, and content reviewed. The learned skills should be practiced. The skills should become part of your routine for learning.
4. Get into the habit of asking and answering questions to guide your study.

QUESTIONS

ADDED CONTENT

- Have you learned how to learn?

- Can you predict what kind of studying you will have to do to learn different types of material?

- What do you do in between tests to monitor your progress in learning the course content?

- Do you make up a graph to depict your grades in various courses? Why or why not?

- Studying and thinking are obviously related. Think about the words representing the concepts below. Match the words in column B to the words in column A.

Column A	Column B
_____ Memorize	1. Create
_____ Understand	2. Evaluate
_____ Use	3. Analyze
_____ Thinking in sequences	4. Apply
_____ Determining right, wrong, fact, fiction	5. Comprehend
_____ Synthesize	6. Know

- Which of the above activities are the easiest or the most difficult?

- How does time relate to these activities?

- Is your study area free from distractions?

6

Study Skills Inventory

Twenty-Seven Study Skills

Check any of the following *study skills* that you regularly use.

_____ Commit yourself to a specific study schedule.

_____ Construct a daily "to do" list.

_____ Construct a semester calendar of course requirements.

_____ Develop an "I will and I can" attitude.

_____ Study science six to ten hours per week, every week.

_____ Preview for lecture and laboratory.

_____ SQ3R text and laboratory manual.

_____ Listen with care and attention.

_____ Take complete and accurate notes.

_____ Get to class on time; quickly review the previous lecture.

_____ Formulate objectives for study sessions.

_____ Review and rework notes frequently.

_____ Write a summary of each lecture.

_____ Outline, chart, or map out the work you must do each week.

_____ Generate questions for self-testing.

_____ Recite, recall, and envision the material studied.

_____ Construct scientific figures.

_____ Practice solving questions and writing brief essays.

_____ Construct a bank of keywords, symbols, and units.

_____ Test yourself with questions from the textbook or manual.

_____ Use a study guide.

_____ Study with a study group.

_____ Review material with a tutor or instructor.

_____ Develop good test-taking skills.

_____ Relate the science you study to the things you read and see.

_____ Analyze mistakes on tests and graded assignments.

_____ Keep up with all assignments; don't cram or procrastinate.

Weekly Time Schedule (chart)

	Sunday	Monday	Tuesday	Wednesday	Thursday	Friday	Saturday
7:00 a.m.							
8:00 a.m.							
9:00 a.m.							
10:00 a.m.							
11:00 a.m.							
12:00 noon							
1:00 p.m.							
2:00 p.m.							
3:00 p.m.							
4:00 p.m.							
5:00 p.m.							
6:00 p.m.							
7:00 p.m.							
8:00 p.m.							
9:00 p.m.							
10:00 p.m.							
11:00 p.m.							
12:00 midnight							

After the first week of school fill in this weekly time schedule. Record times for classes, study, work, recreation, meals, travel, etc. Have you planned 6–10 hours of study for science?

7

Time-Management and Study Sessions

Objectives

1. To recognize the value of establishing specific times to study.
2. To establish a study schedule.
3. To learn what should be done during study sessions throughout the week.

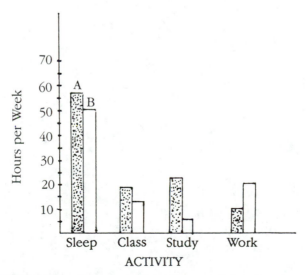

Figure 7.1 Bar graph of weekly use of time for selected activities. Note a *rate* of hours per week is given for each activity.

Q U E S T I O N S	A D D E D C O N T E N T

- How long should the average study session last?

- Imagine that your instructor does not give specific reading assignments. You have your class notes and now you want to check the content of your notes.

  Would you use the table of contents, the glossary, or the index of the text to find the comparable information in your text?

- How would you go about finding a figure in the text that relates to a figure discussed in class?

- Check which of the following would enable you to fill in the gaps left in your lecture notes:
  _____ instructor
  _____ classmate
  _____ lab partner
  _____ study group
  _____ science tutor

- Should you use valuable time to generate and record questions in your notebook? Why or why not?

- Is it worth it to you to summarize notes, formulate information maps, and to devise charts? Why or why not?

Time Management

Time management is an important study skill. Schedule specific times to study science, and stick to that schedule. Schedule six to ten hours *each week*. Study in a variety of ways to learn the information. The study skills inventory lists the many things you can do during these six to ten hours. *It is up to you to schedule and direct your study each week.*

You must tell yourself "I will study science Monday from 8:00 to 9:00, Tuesday from 2:00 until 3:15, Wednesday from 8:00 to 9:30, Thursday from 2:00 until 3:15 and Sunday from 8:00 until 10:00." In addition, you must list the things you want to accomplish during the study session—review notes, make up a vocabulary list, solve certain problems, review the lab, write up answers to the lab questions, and so on.

After all your classes have met for the first time, set up a *semester schedule.* Using a calendar, indicate during which weeks you have tests, when term papers are due, and how often written assignments must be submitted. This will give you an overview of the *tangible* part of the schoolwork to be accomplished. Remember—there is a vast amount of study that is *not tangible;* this is the time involved in surveying, questioning, reading, reciting, and reviewing notes and textbooks. The payback for this time is the gradual comprehension of the information and good performance on quizzes and examinations. Use the weekly schedule form on page 30 to record times you have committed to class, work, meals, travel, or recreation.

On the bar graph, add up the hours you spend each week in the activities represented. Graph your hours next to students A and B. Notice that as you try to account for the 168 hours in the week, many hours sort of just slip by.

The "dead hours" between classes are good times to study. Don't kill these dead hours in the student center. Don't try to study in the cafeteria; studying there doesn't work! It just leads to *R*apid *E*ye *M*ovement, movement of eye from book to passing figures and friends. This form of *REM* generally results in wasted time, frustration, eye strain, and possibly headaches. Use the cafeteria or student center for planned social breaks.

Refer to the "Cycle of Weekly Study" to help you plan how to set up a schedule to study science. Be aware that *self-discipline* and *self-direction* are vital parts to a successful study schedule.

Keep the following in mind as you plan a study schedule:

1. Take control of your time.
2. Schedule enough time to accomplish the task when you are alert and willing to study.
3. Keep the study place free from distractions. Train friends and family not to disturb you, or study where they can't get to you.
4. Schedule study sessions in relatively short blocks of time (40 to 50 minutes) with rewards of five-to-ten-minute breaks. If you can't concentrate for forty minutes, then study for twenty to thirty minutes without a break, but start to train yourself to concentrate for longer periods of time.
5. Arrange the schedule so that your science study will be close to and soon after the time you attend science classes.
6. Set specific goals or objectives for the various blocks of time.
7. Schedule more study time at the beginning of the week than at the end. There is a tendency to goof off toward the end of the week.
8. Use part of a "free day or afternoon" to study.
9. TV is the big "sponge" of time. *Avoid being suckered into watching TV* unless it is on your schedule. *Don't try to study and watch TV at the same time.*

As you study during the week, begin to think about how you can condense and summarize the information you are learning. If you do this, you will have fourteen or fifteen summaries at the end of the semester. Summaries include *outlines* (see the sample outline in chapter 2) and short essays. In addition, *information maps, charts, and tables* allow you to summarize course content efficiently. Summaries will force to choose important information and organize it in a different and personally meaningful way. You will have studied the material you are summarizing and will provide yourself with a useful tool for reviewing the course content.

Q U E S T I O N S	A D D E D C O N T E N T

• What is an example of a "to do" list?

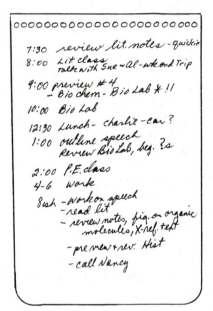

Figure 7.2   A "to do list."

• Give examples of study objectives.

Examples of study objectives are:

1. Learn the structure and function of the cell parts mentioned in lecture and lab.

2. Review notes on moles and molarity. Review sample problems from notes and text. Practice on end of chapter questions.

3. Make up flash cards for the origins and insertions of the muscles mentioned in lab.

4. Construct information table on minerals studied in lab.

10.   Plan time to socialize, goof off, and exercise. These activities will add to your overall well-being.

11.   Plan some time on Friday or Sunday for your end-of-the-week review. This type of study session will enable you to pull together all of the things you studied the previous week. It will also let you preview the coming week. Reviewing and previewing are important study activities.

You can help keep yourself on schedule by making a "to do" list. Use an assignment pad or a pocket calendar for this. An advantage of a list is that it helps you simplify your tasks. The "list" gives you something methodical to do. All you have to say is, "I will do it." The tough part is establishing the skill of composing the "to do" list every day. Resolve to complete the list in the specified time period. Then reward yourself by crossing out the items on the list. Pat yourself on the back.

Most people put off doing certain things because the activities seems overwhelming. Easy things, much lower in priority, are done first. Realize you did the simple things because it felt good to accomplish those simple things. *That is the answer!* Take the *difficult task* of studying what seems like massive amounts of scientific information and *divide it into a series of small tasks.* Follow a system and schedule of study and you will come to grips with the subject matter. The alternative is to procrastinate, let it pile up, and try to learn it all in one cram session. This alternative is *not* the way to learn. It is frustrating and not very rewarding.

You will be expected to learn the science content presented in lectures and laboratories. The majority of your tests (90 to 100%) are based on these sources of information. Textbooks are written for use by different instructors around the country. There is more content in a textbook than you will study in the course. The lecture and lab content is your guide. Comprehend that information, and you will succeed in the course.

Students of the past would advise you to "Stay on top of it . . . don't get behind." For every hour you spend in class, spend at least *one hour in concentrated study outside class.* This time guide presumes you have the prerequisites discussed in chapter 3.

Remember, study involves surveying, reviewing, reciting, recalling, and practicing using the information studied by doing the assigned written work and solving the problems.

QUESTIONS	ADDED CONTENT

- What are information maps?

An information map is a method of organizing information.

A *core concept* or phrase is placed in the center of the page. Categories of information surround the core. Lines connect these major groups to the core. Each category then might be described briefly by short phrases or lists of information. Other subcategories might branch off from these categories that surround the core.

To some, this system does not seem organized. To others, it provides an overall view of a topic. Information mapping might appeal to a "right-brained" individual. (See example in chapter 5.)

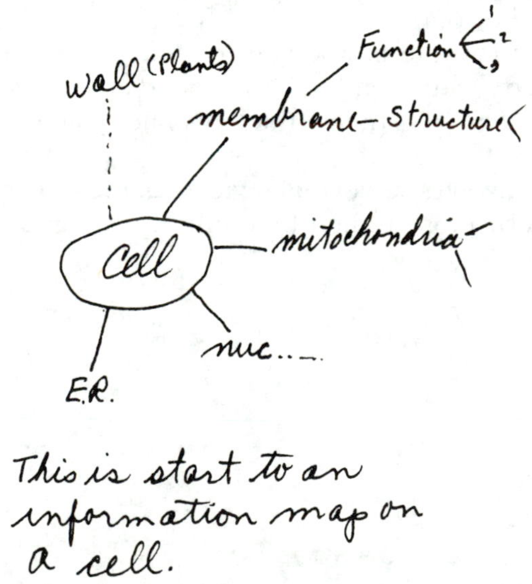

Figure 7.3   An information map . . can you complete the map?

Study sessions

The weekly sequence of classes and study is diagrammed in figure 7.4. You should schedule a preview and review before each lecture. You should certainly prepare for each laboratory. If a recitation class is held, solve the assigned problems before attending class. Note that group study and an end-of-the-week review are included in the schedule.

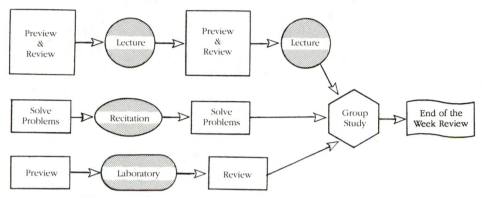

Figure 7.4 Cycle of weekly classes and study sessions. This is an information flow chart. The beginning of the week is to the left, the end to the right.

Preview for Lecture

If the instructor has given definite reading assignments, then you can preview the lecture material. The objective of the preview is to *develop a general idea* about the content of the next class. Don't feel you have to learn it all during the preview. All you have to do is

- survey the chapter objectives, headings, subheadings, and highlighted words
- read just the introduction and chapter summary;
- survey the figures in the chapter;
- generate questions about the previewed material and read the related questions at the end of the chapter;
- try to predict what will be covered in class.

QUESTIONS

ADDED CONTENT

• What is the important material to study?

Chapter 2 contains a list of suggestions to help you identify the important material to learn in a science course.

Summarize those suggestions here. To do this, use a short essay, a list, flow chart, or an information map.

• What are the different study activities that will help you learn the material?

• What is the difference between previewing and reviewing?

Review the checklist of study skills on page 29 and the information maps in chapter 5.

• Why should you preview material before going to lecture or laboratory?

• If you follow the recommended study schedule, how many different science study sessions will you have each week?

Preview for Laboratory

When coming to lab, one student often asks another in a supposedly "cool" manner: "Hey, what are we doing in lab today?" It might sound cool, but it's *dumb*. Avoid this negative behavior!

A positive approach would be to preview the assigned exercises during a study session. You should become familiar with the terms, concepts, and procedures of the laboratory exercises.

- Survey the entire exercise first. Read the objectives carefully. Get a general idea of what is in the introduction and procedures. Survey any questions in the exercise.
- Generate your own questions about the objectives and procedures. Answer these along with the questions contained in the exercise after you have completed the lab.
- After doing this, read the introduction more carefully. Relate the content to the content of the lecture.
- Next read through the procedure. Underline important parts. If you really want to be prepared, make up flow charts or diagrams of the procedures. Label the amounts and types of materials to be used. By doing a thorough preview, the mechanics of the laboratory will be less intimidating. You will have time to think about "what you are doing in a particular exercise" and "why" you are doing it.

Before-Class Review

Get to class early and skim through your notes. Compare your notes with those of a classmate. Talk about the subject matter for a few moments rather than "throwing the bull." Recall and recite the material you previewed.

Listen to the instructor's review. Put asterisks or exclamation points next to the things in your notes the instructor mentions.

End of Class

Listen for the reviews and previews the instructor gives near the end of class. Take notes on what is said! They will help guide your study.

QUESTIONS

ADDED CONTENT

- What should you study to prepare for the lab?

When you study the laboratory material, you should

- know the names and functions of the tools and apparatus used;

- learn the names and function of the materials used;

- remember and understand the results of experiments;

- memorize the units of measurements and symbols of the terms and concepts;

- be able to interpret maps, graphs, instrument readings, and reagent test results;

- be able to describe and discuss what relationships you verified in each exercise or experiment.

- How should you study information covered in lab?

Gather the information, record results, and take complete notes. Make an effort to diagram the procedures and equipment. *It is important that you can visualize what you did.* Diagrams, no matter how crude, will help you visualize the experiments and that will help you learn.

All that remains is the need to correlate the terms, units, reagents, materials, and *concepts.*

The questions found at the end-of-the-laboratory exercise will test your ability to correlate information and experimentation.

Answer the questions.

Review of Lecture

Long-term memory increases when you review material frequently. *If you don't use it, you'll lose it.* Early review of notes will enable you to identify difficult areas. You will then have time to work out the answers or seek help from your instructor or science tutor. During the review phase of study, you should do the following:

- Review the lecture notes as soon as possible (within 24 hours). *Highlight* or *underline* keywords.
- Note gaps. Fill these in or make a note to clarify information with instructor.
- Write a brief summary in the space at the beginning of the day's lecture notes.
- *Analyze any diagrams* in your notes. Compare these to those that appear in the text. Make any *corrections* or *additions*. Redraw the figure in the added content column. Explain the details of the diagram to a pet, peer, or friend.
- *Compare* the definitions and examples given in lecture with those given in the text. Make corrections.
- Make up flash cards.
- *Read* pertinent *parts* of the *text* (SQ3R method is recommended). Take *notes on material* that will *enhance* your learning. Record these in the added content column. Remember—you should use your class notes to identify what areas of the text should be read.
- Identify the material covered in the lecture and the text. If it is found in both places, then it is information you must learn.
- Compose information maps or charts when appropriate.

Study Session for Laboratory

Complete the study of your laboratory work within two days after finishing the lab. If you have difficulty with the lab write-up or questions, then you will have plenty of time to get the answers before the next lab period. Do the following:

- Review the objectives, introduction, and procedures.
- Clarify and correct notes or data taken down.
- Analyze the data and observations; draw conclusions and answer the questions at the end-of-the-lab exercise. Answer the questions you might have generated.
- Make a note of any unanswered questions.
- After this is done, visualize what you did, what you observed, and what your results were.
- Be able to describe the procedures and the materials.
- Relate the laboratory information to the lecture content.

Q U E S T I O N S	A D D E D C O N T E N T

• What is an information table?

Information tables

An information table is a system that organizes information into rows and columns. The organization allows you to compartmentalize information, but like the information map, it *requires* you to make the interpretive connections between the compartments.

This system of review is rigid, concise, and concrete. It probably would be very appealing to you if you were a "left-brained" individual.

Extraembryonic Membranes in Chick and Human

Name	Germ Layers	Location in Chick	Function in Chick	Location in Humans	Function in Humans
Chorion	Outer layer of ectoderm and inner layer of mesoderm	Lies next to shell	Gas exchange	Fetal half of placenta	Exchange with mother's blood
Amnion	Outer layer of mesoderm and inner layer of ectoderm	Surrounds embryo	Protection; prevention of desiccation	Same	Same
Allantois	Outer layer of mesoderm and inner layer of endoderm	Outgrowth of hindgut	Collection of nitrogenous waste	Same	Blood vessels become umbilical blood vessels
Yolk sac	Outer layer of mesoderm and inner layer of endoderm	Outgrowth of midgut; surrounds yolk	Provision of nourishment	Same, but contains no yolk	First site of blood cell formation

Figure 7.5   An example of an information table used in a biology textbook. If your instructor displays this kind of table and lectures from it then you must remember the information. You must learn it.
(From Mader, *Inquiry into Life,* 6 ed., Wm. C. Brown Publishers, Dubuque, IA)

Laboratory Books and Reports

Courses for science majors might require the recording of procedures and data in *bound* lab notebooks and the writing of formal laboratory reports. The actual requirements vary from college to college. You should be sure to know what *your instructor's requirements are and adhere to them.* Careful and organized record-keeping is part of the laboratory exercises. These requirements are reviewed during the first laboratory period. If the requirements are not clear, then ask questions. Clarifying what is expected is not "stupid"; it is *smart.* The following are things to keep in mind:

- Observations and data should be recorded neatly.
- Graphs, tables, and diagrams should be clearly *labeled with captions and units.*
- All calculations must be shown.
- Incorrect data or calculations should be crossed out with one line. Don't tear out pages or blot out work.
- Written work (analysis and conclusions) should be concise.
- Hand work in on time; don't let work pile up.

Recitation Classes and Study Sessions

The problem-solving classes relate the principles and laws discussed in lecture and laboratory. Sets of problems are assigned and then reviewed. The textbook is generally the source of these questions. Chemistry and physics books give sample solutions to the different types of problems presented in the chapter. The questions at the end of the chapter are generally arranged in the same sequence as the material is presented in the chapter. *It is vital* that you practice solving problems on your own. *No pain, no gain.* Do the assigned problems. If you need more practice, do the unassigned problems. Work until you are confident that you can solve the problems related to the topics discussed in lecture. Remember—the lecture is your guide. It is the problem-solving process that helps you apply the principles and laws.

After you attend a recitation class, review the solutions discussed in class. Make sure you have no lingering doubts. If you do, go to the instructor or tutors for help. *After completing your review, start the next set of assigned problems.* Chapter 12 suggests and discusses problem-solving techniques.

QUESTIONS   ADDED   CONTENT

- What is an example of an information map?

- Could you write an essay using the information in this map? Try to do it or explain the map to a friend.

Energy Map –

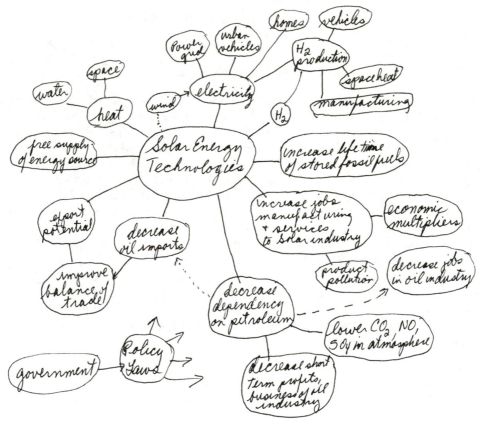

Figure 7.6   This is an information map. Start at the center and follow the lines out to subcategories of information.

End-of-the-Week-Review

During the end-of-the-week review, recall and integrate the material presented in class. Summarize the week's work into outlines, information maps, or charts and essays. After you do this, then you should test yourself. How well do you know the material? Make up your own test. The questions for this self-testing can come from a number of sources:

- Your notes should have questions recorded during the week of study that either you or your instructor asked.
- You can make up questions from objectives listed in your textbook or laboratory manual.
- The text and manual have questions at the end of the chapter or exercise.
- The chapter headings and subheadings can be rephrased into questions.
- You can purchase a study guide.
- Students can exchange questions to test each other.
- Learning centers might have computer-based study guides and tests.

Group Study Session

An effective study group can help pull together the information presented each week. Three to six students form a good working group. You should meet for at least two hours with the intent to study science, not to discuss personal or world problems. Work with people with similar goals and objectives; avoid academic parasites. It is important that each member of the group prepare for the session by reviewing, correcting, and clarifying his or her own notes. Questions should be generated and problem-solving attempted.

The group must decide on a sequence of study, and stick to it. Certainly you must relax and enjoy one another's efforts, but resist any long-lasting distractions to the group effort. This is the time to practice what you have learned. Your study group might do the following:

- Compare notes. One person should recite the notes while the others listen and make additions or clarifications. This will be slow going at first, but as the semester progresses it will get easier as everyone's note-taking skills improve.
- Analyze any figures from the lecture, text, or manual. Have one person explain the figure.
- Help each other answer the questions from the various sources of questions (see above).
- Review the textbook to highlight areas covered in lecture or lab.
- Help each other form information charts or maps.
- Test each other.

- What are examples of information tables?

Physics - information tables can be used to summarize concepts.

Name	relationship	symbols	description
Second Law of motion	$F = MA$	F = force (N) M = mass (kg) A = acceleration m/s² N = newtons kg = kilogram m/s² = meters per sec squared	The net force acting upon a body is equal to the product of the mass and the acceleration of the body. The unit of force is the <u>newton</u>
Third Law of Motion	$F_{AB} = -F_{BA}$	F = force A B interacting bodies	When a body (A) exerts a force on another body (B) the second body (B) exerts a force on the first body (A) of the same magnitude but in the opp. direction

Chemistry - an information table:

Gas Laws:

Name	relationship	symbols	description	application
Boyle's	$PV = k \left(\frac{constant}{T}\right)$	P = pressure V = volume k = constant i = initial f = final	At constant temp., the pressure and vol. of a sample of gas are <u>inversely</u> proportionale	If conditions on sample change then $P_i V_i = P_f V_f$
Charles	$V = kT$ (constant press.)	V = volume T = temp. (K) k = constant i = initial f = final	At a constant pressure the volume of a gas is <u>directly</u> proportional to its absolute temp.	If either the V or T conditions change then $\dfrac{V_i}{T_i} = \dfrac{V_f}{V_f}$

Figure 7.7   Information tables are valuable ways to summarize content of courses.

- If you are taking a problem-solving course, be sure to work on problems as a group. One person should explain the steps to a solution. Help each other, but don't get impatient. Check to make sure that everyone in the group understands the process of the solution.
- Each person in the group should vocalize some part of the group review.
- *Thank one another for academic and moral support.*

Review

1. Establish a fixed weekly cycle of study and stick to it. Your success depends on it.
2. Study in a place where you will not be distracted.
3. Compose "to do" lists and establish learning objectives.
4. Divide big tasks into smaller, simpler ones.
5. Preview text and previous lectures notes before attending lecture.
6. Be an active listener and note-taker. Be an active experimenter in lab.
7. Review and revise the lecture notes as soon as possible after each lecture.
8. Recall and recite the lecture content several times each week.
9. Preview the laboratory exercise before attending the laboratory.
10. Practice problem-solving continually if you are taking a problem-solving course.
11. Generate questions for lecture and lab material. Use these questions to continually test yourself.
12. Once each week, integrate and recall the week's material in a review session.
13. Study with other students in productive study groups.
14. Stay on top of the material and don't procrastinate.

Class Notes Checklist

Use this checklist to judge the quality of your class notes. Check "Yes" or "No" if these are included in your class notes.

		YES	NO
1.	Date and subject of lecture	____	____
2.	Recording of assignments	____	____
3.	Summary statement of lecture	____	____
4.	Format		
	a. consistency of style	____	____
	b. phrases rather than full sentences	____	____
	c. outline form with indentions	____	____
5.	Neat and legible	____	____
6.	Accurate (check with instructor or classmate)	____	____
7.	Clear, can be remembered	____	____
8.	Complete		
	a. record of instructor's hints and gestures	____	____
	b. record of what is discussed as well as what is written on the board	____	____
	c. diagrams recorded in large size and appropriately labeled	____	____
	d. references to diagrams not recorded in notes	____	____
	e. record of examples of solutions to problems	____	____
9.	Use of symbols and abbreviations		
10.	Starring or marking of important points	____	____
11.	Leaving spaces for points missed or points of confusion	____	____
12.	Notes taken on the right-hand page	____	____
13.	Review and clarification of notes within four hours of the class	____	____
14.	Recording of questions in the questions column	____	____
15.	Recording of instructor's questions	____	____
16.	Comparison of the lecture content with the textbook content and identification of what is covered in both places	____	____
17.	Outline of pertinent textbook content in added content column	____	____
18.	Summation of week's work in outline, essay, or map	____	____

CHAPTER

8

Class Notes

Objectives

1. To learn the ingredients of good listening.
2. To learn how to organize a notebook to make it a useful learning tool.
3. To learn the characteristics of good note-taking.

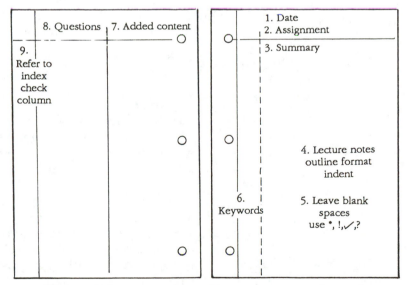

Figure 8.1   Organization of lecture notes. Create a tool rather than a "jumble." The numbered parts are explained in the text.

Q U E S T I O N S

- What is involved in good listening?

- Have you trained yourself to listen carefully?

- What can you do if your listening skills are poorly developed?

- How do the recommendations suggested on note-taking differ from your note-taking style?

- Can "old dogs" learn new tricks?

- Which method of note-taking gives you the most complete notes?

- Which note-taking style is more work?

- Why leave blank areas in your notes during the lecture?

- Can you do a better job taking notes? What are you going to do to take better notes?

- How can you judge the quality of your notes?

- Compare notes with a classmate. Discuss any differences. Whose notes are better?

A D D E D   C O N T E N T

When listening:

1. Don't be judgmental as you listen. Your objective is to learn the content and concepts of the presentation.
2. Enjoy a good delivery, but learn to tolerate a poorly delivered lecture. Don't forget to take notes in either type of lecture.
3. Don't get distracted by repeated "Uuuh's," "Okays," and "You knows."
4. A brief preview will help prepare you to listen better. Review your notes and preview your text.
5. Be willing to change old concepts and attitudes.
6. Keep your mind open to new ideas and concepts.
7. Realize that college science will increase the depth and breath of knowledge you already have.
8. Try to tune out distractions . . . stray thoughts, problems, anxieties. This is difficult to do but very important!
9. Keep trying to extract the broad concepts, then fortify these with specific information.
10. Be attuned to the gestures, tone, and body language of the instructor.
11. Be physically and mentally alert.

Remember, it is important to keep alert during lecture.

If you take notes, record keywords, and record questions, then your mind will be constantly busy, constantly working. You can't help but be alert. You will naturally start to become a better listener and an active learner in lecture!

You should become confident that you're starting to do the things necessary to become a successful student. If you're taking good notes, you will be providing yourself with a valuable learning tool. If you are doing that, then give yourself a pat on the back.

Notes

Notes taken in lecture and laboratory are the *primary guide* to what you should learn. If you have not recorded what is covered in class, then your studying will lack direction. Studying will be difficult and frustrating. Notes should be taken on the material written on the board *and* on the material discussed by the instructor. Even though you say you will remember what was covered in class, experience and experiments indicate that you will *not. Remember,* 90 to 100 percent of science tests are based on material covered in class.

If you have had difficulty learning science in the past, then one of your problems might have been poor note-taking skills. You should decide to develop and improve this skill. It takes time, work, and practice. If you do improve your note-taking skills and if you use these improved notes effectively, then you will increase your chances of successfully learning science. You will understand the course material. *You will pass!*

Additions to Class Notes

During the end-of-the-week review, you should outline or write a brief summary of the week's work. Construct information charts, tables, or maps. Reorganizing and condensing the week's material will enable you to study the course content in yet another way. These materials should be added to your notebook. By the end of the semester, you will have fourteen or fifteen packets of weekly reviews that will make studying for the final examination easier.

You might also want to consider adding outlines from specific sections of the book to your notebook. Try to put these next to the class discussion of the same material. Keep similar material together.

Listening

Listening, observing, and writing are three class activities that reinforce each other to allow you to learn the material more easily.

A good listener

1. pays attention to what the instructor is saying
   (don't let your mind wander).
2. slows down the pace of thinking to the instructor's rate of speaking.
   (Thinking occurs at the rate of about 400 words per minute; most instructors speak at about 100 words per minute.)
3. accepts the instructor for what the instructor is and follows his or her train of thought.
4. practices the skill of listening and commits what is heard to memory.

Realize that if you become a good listener, then class time will actually be "learning time." It is your responsibility to listen in class. College instructors do not necessarily feel that it is their responsibility to *get you to listen.* Of course, many instructors do use humor, odd facts, and interesting tales to attempt to motivate students to listen.

- How soon do you start forgetting things you have learned?

- Why is it that you don't forget the information you constantly use?

- How will frequent reviews change the amount of material you forget or remember?

- Why will studying the same material in different ways tend to help you remember it?

- What is the difference between memorizing something and understanding something?

- Which type of material do you tend to forget more rapidly: the material memorized or the material understood?

- *Volume, velocity,* and *vortex* are words you could memorize along with their meanings, but would that mean you understand the terms and how they can be applied?

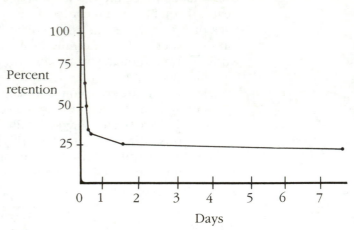

Figure 8.2   Curve of forgetting.
Source: Adapted from Annis, L. F. *Study Techniques.*
Dubuque, Iowa: Wm. C. Brown, 1983.

You will forget about 60 to 70 percent of the material you hear in lecture within 24 hours. If you want to remember it, then you must review it frequently.

A remembering curve with frequent reviews would look something like the one in figure 8.3.

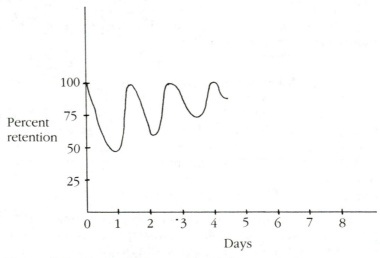

Figure 8.3   Remembering curve with frequent reviews.

Active note-taking helps keep you alert. Recording notes helps transfer the oral word to the written word. Seeing the written words provides for visual learning. Thus, careful note-taking will in fact be a source of three types of stimuli: auditory, visual, and tactile (the feel of writing and forming the notes).

Notebook

Use a standard-size looseleaf or spiral notebook. A looseleaf is preferred because new pages, handouts, and returned tests can be added to the notebook. The notebook should be organized as shown in figure 8.1.

Right-hand Page

The following numbers refer to the numbers in figure 8.1.

1. Each lecture should be dated and titled.
2. Assignments should be recorded.
3. Several lines should be left blank. When you review the notes for the first time, write a short summary of the lecture in this space.
4. The bulk of the right-hand page should be used to record your notes. This should be done in simple phrases or in outline form. How you do this depends on your present note-taking style and your instructor's teaching style. Indentions of less important material helps separate these points from major ideas.

   • Dashed lines and numbers or letters also help.
   • Be consistent in your style.
   • Write legibly.

Leave a space if you miss something or find something confusing. Then when you review your notes, you'll know where you had a problem in lecture. You'll also have the space to clarify the notes.

* *In the spaces left blank, you will be able to correct spelling, add or correct information. If you don't leave spaces like this, you'll have to squeeze in corrections... or worse, never make changes. Leave spaces — Paper is Cheap!*

QUESTIONS

- Why go to the trouble of making up questions?

- Do you think it is a good idea to divide the left-hand page into "questions" and "added content" columns before lecture? Why or why not?

- Why is it important to record the questions the instructor asks?

- Can you possibly learn to take notes, record keywords, and record questions during the lecture?

- How do you make up questions?

ADDED CONTENT

By now you have noticed that all introductory science books have chapters divided into sections. Each section has *headings* and *subheadings*. Keywords are often in bold print.

The headings, subheadings, and keywords can all be converted to questions. Simply place the word *how, when, where, what, explain, discuss,* or *compare* in front of the title, subtitle, or keywords.

The heading "ATOM" becomes "What is an atom?" or "What is the structure of an atom?"

The heading "Molecule" becomes "What is a molecule?" or "Explain the different kinds of molecules discussed in class."

Don't feel you can't make up questions. You can! As time goes on, you'll improve your questioning skills.

Most importantly, you will become better at predicting what questions your instructor will ask. This takes practice.

Ask your instructor to check over the questions you ask.

Compare the questions you make up to the questions your instructor asks on tests.

5. Use — , * , ? , and ! to identify important points. When your instructor begins or ends lecture, he or she might summarize material. Don't just sit and listen; go back to your notes and * or ! the points reviewed in the summary. If you can't find the material summarized, then take notes on the reviewed material. If you don't do this, you're giving your mind a chance to wander off in a flight to fantasyland. *Don't close down your listening and note-taking until your instructor stops for the day.*

6. The margin of the right-hand page should be reserved to record keywords, new words, or what might really be called *test words.* During your reviews, these words should be incorporated into a vocabulary list or flash card system. If you find the margin too narrow, draw in a wider column to match your needs. *Remember, these keywords are potential test words.*

Left-hand Page

7. At least half of the left-hand page should be used to

- add to or clarify the content of class notes;
- correct or redraw diagrams;
- record information maps or summary tables;
- notes from the textbook that relate to the lecture.

8. Save at least one third of the left-hand page for a *question column* (review figure 8.1). You should continually make up and record questions that pertain to the information in your notes in this column. *Asking questions will*

- help you focus on the material to be studied;
- enable you to test yourself before the instructor tests you;
- indirectly improve your note-taking and listening skills;
- make you an active learner by making you a questioning listener.

Remember, questions asked should be written in full sentences. Ask your instructor to evaluate them. Ask not only *what* but also *how, when, where,* and *why.* Avoid asking questions with a yes or no answer. If you do, then follow these with a "Why" or "Give the reasons for."

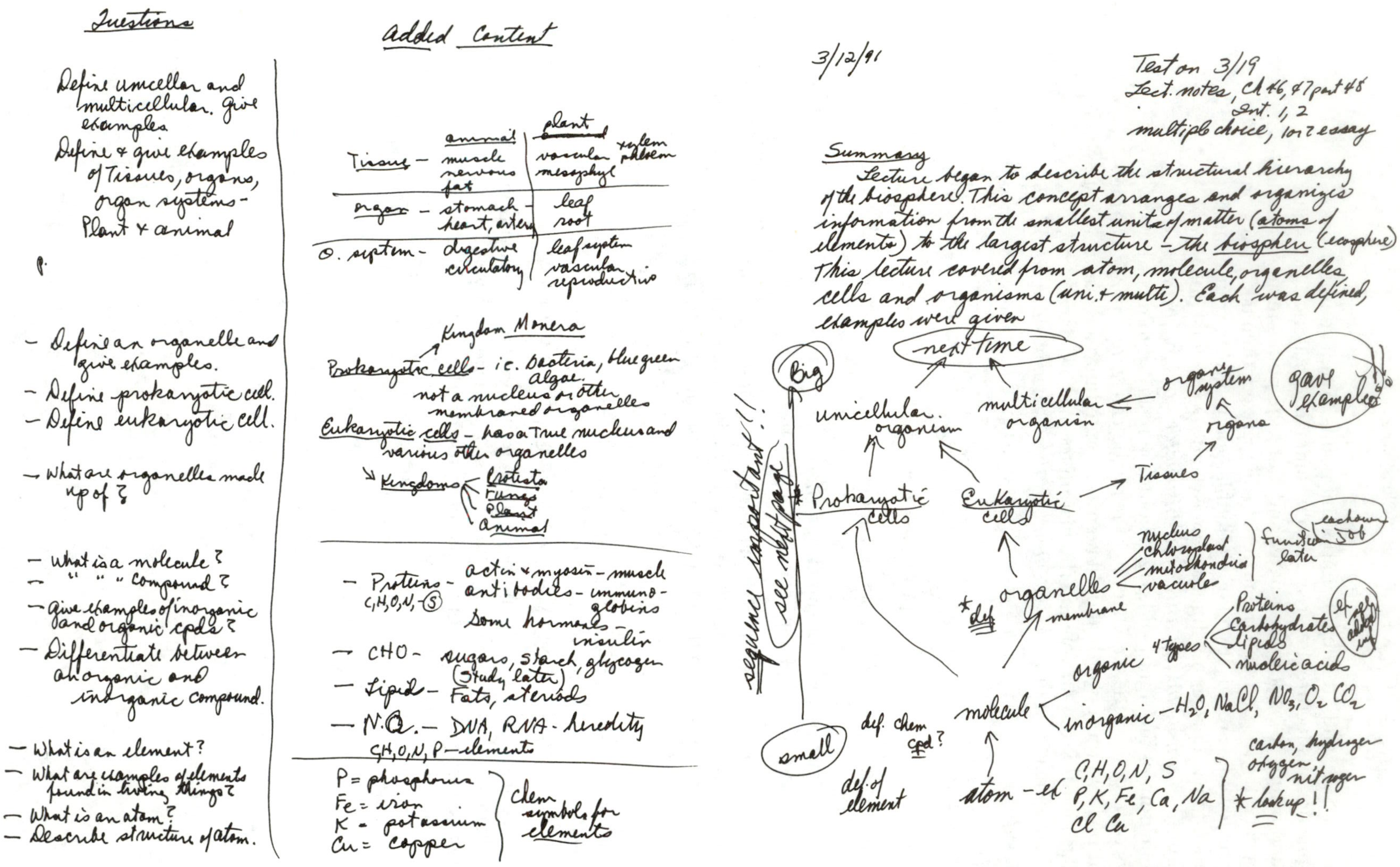

Figure 8.4a This is an example of a good set of notes. The information reflects not only the blackboard information but also verbal information. The information is more complete and accurate. The *Question* and *Added Content* page and circled notes direct student study and learning.

Making up many questions will do you no good if you don't use them. During the end-of-the-week review or in your study group, ask and then answer the questions. Cover up the notes, read the questions, recall and recite the answers. Check the notes to see whether your answers are correct.

> ******** *After the first test, compare your questions to the test questions.* How many were helpful? Refine your questioning skill. *It will pay off!*

9. The margin of the left-hand page might be used for a number of things:

- Place a check here if you passed a self-test.
- Record pages of figures in the text that help answer the questions.
- Place a question mark here to indicate that you should ask your instructor this question.
- Record the page in your text where content of lecture can be found (search of index).

Additional Hints

1. Develop your own shorthand. Eliminate vowels and use symbols. *Develop* becomes *dvlp; and* becomes *&; turns to* becomes ——->.
2. Use tape recordings only to check or clarify notes, not to listen to the whole lecture again.
3. Note whether your instructor repeats material in different ways or gives different examples concerning the same topic.
4. Record clues given by the instructor. The tone of voice, the use of expressions, and various gestures are all important parts of body language. Use !! or ** to record nonverbal means of emphasis.
5. Be alert for *indicator words or phrases.* Examples:

- *There are three important . . .*
- *The beauty of this is . . .*
- *Therefore,* or *In summary . . .*

6. Attend all lectures. Don't depend on other people's notes unless they are as good as yours.

> (A Cornell University student started a note-taking business to provide quality notes to more than 3,500 subscribers. Since people are willing to pay for good notes, they must be important.)

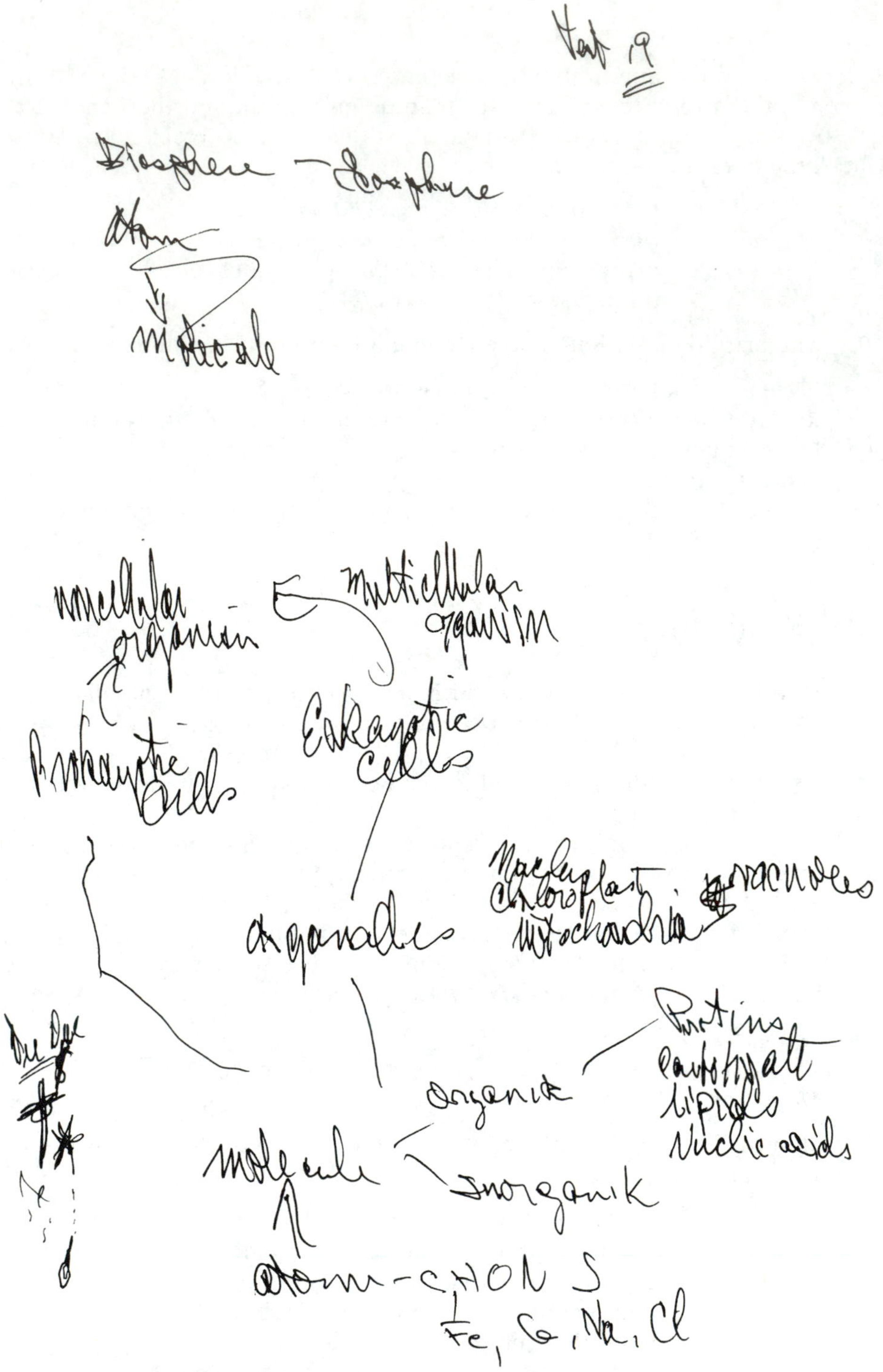

(handwritten notes)

Biosphere — Biosphere

Atom

Molecule

unicellular organism — Multicellular organism

Prokaryotic Cells — Eukaryotic cells

organelles — Nucleoplast chloroplast vacuoles mitochondria

organic — Proteins carbohydrate lipids nucleic acids

molecule — Inorganic

atom — CHON S
Fe, Ca, Na, Cl

Figure 8.4b This is an example of poor note taking. The information in the notes are what appeared on the blackboard. The information is incomplete and not accurate, full of errors.

7. Be sure to take notes on the discussion as well as on what is written on the board. Record the examples given to demonstrate the concepts.
8. If figures are shown on transparencies or slides, make as quick and simple a diagram as possible. Search for similar diagrams in your textbook. Enrich and correct your diagrams during review study sessions.
9. Be sure to copy diagrams from the board accurately!
10. Give some indication in your notes as to how much time the instructor spent on a topic.
11. Preview textbook material and the previous day's notes before attending the next class.
12. Review and revise class notes as soon after lecture as possible.
13. Keep notes simple but complete.

Review

1. Attend all classes.
2. Take notes on the right-hand page.
3. Record keywords (test words) in the margin of the right-hand page.
4. Use your own shorthand.
5. Keep mentally active in lecture.
6. Generate questions to test yourself.
7. Revise, review, and recite your notes frequently.
8. Compare notes to the textbook content.
9. Be sure to study and comprehend all figures discussed in class.
10. Remember, 90 to 100 percent of test material comes from class material.
11. Test yourself with the questions you have generated.
12. Compare the questions you generate with your instructor's test questions.
13. Practice and improve your listening and note-taking skills.
14. Review notes with other students.

QUESTIONS

- Which of the textbook characteristics listed in the added content column are in your text?

- Which two can help you locate a topic in the text?

- Which one can help you define a specific term?

- Which three can be used to generate your own questions?

- Why do certain words in your text appear in bold print?

- Most textbooks contain more information than is presented in the course. How can you select the material that you should study?

- Which symbols used in the text should you learn?

- Why is it important to learn the symbols?

ADDED CONTENT

Textbook Checklist:

Compare the table of contents with the course outline. As you do this, have a pencil in hand to help you focus on the material you read.

After doing this, thumb through the book to determine its organization.

Check which of the following are included in your book.

_____ Chapter titles
_____ Chapter subtitles
_____ Chapter objectives
_____ Bold printing of keywords
_____ Keyword list
_____ Introduction to each chapter
_____ Figures within chapters
_____ Margin comments
_____ Portions in boldface print
_____ Boxed discussions or examples
_____ Chapter summary
_____ Review questions
_____ Examples of solutions to problems
_____ "Terms to remember" lists
_____ Use of symbols
_____ Use of mathematical equations
_____ Answer keys
_____ Chapter test
_____ Glossary
_____ Index

Think of how you should use the publisher's organization of the text to help you study effectively and efficiently.

9

Use of Textbooks

Objectives

1. To realize that textbooks provide information to reinforce the lecture and laboratory content.
2. To understand that textbooks must be studied slowly and carefully (pencil in hand), with specific learning objectives in mind. The text is designed to help you learn the material.
3. To learn a technique to extract information from the textbook or laboratory manual.

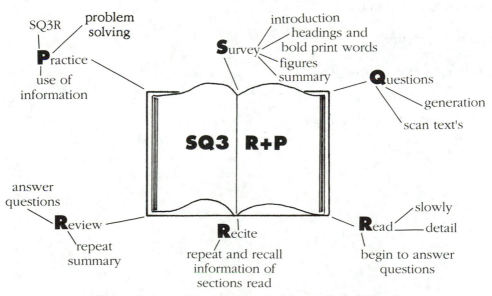

Figure 9.1   SQ3P plus P method for reading and study. This is an example of an information map.

Q U E S T I O N S	A D D E D C O N T E N T

- What determines the information you should learn?

- What kind of lecture style does your instructor have?

- How similar is the lecture material to the material presented in the text? The keywords are the same, but what about the examples and figures?

- How can you determine which material to concentrate on in your text?

- What is a learning objective?

- Do you think it is worth your time to list a series of objectives? Why or why not?

- What does SQ3R plus P mean?

- Explain what is suggested to be accomplished in each step of SQ3R plus P.

- Explain why SQ3R does or doesn't make sense.

- How does this method of text use differ from the way you have used textbooks in the past?

- Why should you have a pencil in hand?

- Does it make sense to practice?

- How do you plan to put SQ3R plus P into practice?

- How will you incorporate this study skill into your *cycle of weekly study*?

The lecture notes, textbook, and laboratory exercises are sources of information you will study to learn science. *Experience demonstrates* that the required course *content* is *set* by the lecture and laboratory *instructors.* Thus, the primary guide to your study is the content of *complete notes from lecture and laboratory classes.* The textbook and laboratory manual will generally be used to clarify, reinforce, and supplement what is covered in classes. Follow the instructor's lead to the depth and breadth of the course content. Compare the lecture content to the textbook content. Concentrate on the information mentioned in both places.

Textbook Organization

Believe it or not, a textbook is organized to assist students to learn. The information is challenging, but publishers do make an effort to break the subject matter into manageable segments. Compare your text to the textbook checklist on page 60. *Headings, subheadings, and bold print* identify keywords or concepts. *Introductions and summaries* to chapters survey the information. *Figures* summarize and help present the subject visually. Samples to *problem-solving* are given. End-of-the-*chapter questions and sample tests* allow for self-testing. The *glossary* is a word list, and the *index* helps you find things.

Reading Skills

As you use the text, you must be able to

- read flexibly, for detail or for general concepts;
- learn from the various types of figures and be able to understand the symbols of science;
- read sequences of directions accurately and carefully;
- generate and interpret questions;
- analyze and evaluate data and information;
- draw conclusions based on the analysis;
- apply the information in critical thinking about everyday or scientific problems.

Reading Objectives

You must have an objective in mind when you use the text, and you must use a consistent technique to fulfill that objective. Don't just read. Science books are not novels. It would help you focus your study if you actually made up a list of objectives. The following are examples of learning objectives:

- *I am going to learn about* the function of the pituitary gland, the nature of a cold front, the concept of neutralization.
- *I am going to compare* metamorphic and igneous rock, mass and weight, velocity and acceleration.
- *I will distinguish between* covalent and ionic bonds, mitosis and meiosis, conduction, convection, and radiation.
- *I will compare the content of today's class notes to the content in my textbook.*

- How much time should you spend surveying the information in the textbook?

- Why is it that one person can "study" a textbook for hours but not learn a thing?

Not very much. Surveying should be a short burst of concentrated study; ten to fifteen minutes should do it. Remember, a quick overview of the topic is all you want to do.

A survey is meant to familiarize you with the material, to warm up your mind to the topic. You are essentially "stretching out" or "warming up" to reveal what objective you should accomplish during the Q3R segment of studying the text.

Make a "how to survey" card, listing the surveying activities on a 3-by-5-inch card. Keep this card in your text as a bookmark. If you should forget what to do when surveying the text, refer to the list. Keep reminding yourself what a good surveying technique is until you have it down pat.

Don't start to survey and read the whole chapter if you will only be covering one third of the chapter in lecture.

Survey, Question, Read, Recite, Review, and Practice

The chapter assignments, the lecture content, and lab exercises will help you locate the material to be studied in your textbook. Keywords recorded in the margin of your notes can be looked up in the glossary and index. *Again . . .* if the information is mentioned in the lecture and the text, then *this is the material you should be sure to learn.* It is prime test material!

Knowing where to find the material is the first step. Defining learning objectives is the second. The third step is to learn the material you are studying. How do you get it out of the book and into your thick head? If you're going to try to do some serious studying, to use your time effectively and efficiently, then choose the proper environment. Reduce distractions; don't play your favorite music, don't have a picture of your boyfriend or girlfriend in front of you, don't have the TV on, don't try to study in busy areas of the library or student center or in the cafeteria. Yes, you can study with all these distractions, but it will not be as efficient; various stimuli will be competing to get into your cerebral cortex.

Adopt a method to study your textbooks. The SQ3R plus P method of study is recommended. Remember, you should follow a concrete, logical method of study to fulfill the learning objectives you have listed. Avoid the predicament of "I studied the text for hours; I can't understand why I got a 52. I must have overstudied. It's not fair. Gary got an 84 and hardly studied."

Survey

Read the introduction to the chapter, the list of objectives if present, and just the headings and subheadings. Scan the words in bold print and the summary. Do not read the text of the chapter. Read the captions and examine the figures. Develop a general view of the chapter or the section of the chapter you are surveying. Scan the questions or problems at the end of the chapter.

Question

Generate questions from the headings and subheadings of material surveyed. Record these in the questions column in your notebook. Reread the questions you have generated in your lecture class. You will begin to have an idea of what to study. These questions can serve as learning objectives.

- How should you read the content of a chapter in a science textbook?

You should read slowly and carefully. You might find it helpful to guide yourself through the text with a pencil. In addition, you might find it helpful and natural to read the difficult parts out loud or in a whisper. Verbalization and pencil tracking seem to make difficult sections seem somewhat easier. This technique is particularly helpful when you are studying sample problems in physics and chemistry textbooks.

In addition to this careful and detailed (every word counts) study, you should stop every once in a while to cross-reference the information you are reading with the information in your notes. How did the instructor teach the material you are reading? All of this takes time and effort. That is why six to ten hours of study is recommended in the cycle of weekly study. Read and study to answer questions and reinforce the content of lecture and laboratories.

Read

With pencil in hand, slowly and carefully read to find the answers to the questions you have generated. You might want to underline, star, or mark certain phrases. *Avoid underlining entire paragraphs.* By using pencil, by answering questions, and by fulfilling learning objectives, you will be a more active and focused reader. Of course, all of this takes time.

Recite (Recall)

Before racing to "finish off the chapter," you should pause to recite and recall the answers to the questions for particular sections of work. Recite the equations used and explain the relationships represented in the equations. Try to envision the figures presented or the example problems. Pat yourself on the back if you recited correctly. Go back and reread if you stumbled or drew a visual blank.

Review

When you finish several sections or a chapter, then review the material. Repeat the survey. Go over the headings and the subheadings, keywords, captions, figures, and summary. This is a good time to compare your lecture notes and the textbook material. Construct information maps or charts to summarize what you have learned. If the instructor did not talk about or assign a specific section of the chapter, then this should not be part of the learning objectives list. However, if you're interested in that material, then read it.

Practice

Practice using the material you have learned. Try to apply the information to the world around you. This will be easier for some courses than for others. Read articles in papers and magazines that relate to your study. Explain what you have learned to classmates, parents, friends, or pets. Draw figures about the material and explain them. Practice solving assigned and unassigned problems in a problem-solving course. Problems will surely be on the tests. *You can't wish your way through them.* Most people will agree that "If you don't use it, you'll lose it." If you use what you've learned, you'll retain it.

Q U E S T I O N S	A D D E D C O N T E N T

- How can you use the textbook headings and subheadings?

- Do the headings and subheadings form a topic outline of the chapter's content? How can you use this information to help you study?

- Why should you compare the content of your lecture notes to the headings and subheadings in your textbook?

- What do you do when you read a word you do not understand?

- Do you see all the words and symbols that appear in your notes and textbook, or do you just skim over them?

- When you use your textbook, how do you determine what to study?

- What is the lecture style of your instructor? (See chapter 2.)

Laboratory Manual

As you preview and review each laboratory exercise, use SQ3R.

Compare the organization of the laboratory exercises to the textbook checklist. You will see similarities and differences. As a general rule, exercises have an introduction, a statement of objectives or problems, list of materials, a sequence of procedures, and a place to record observations, analyses, and conclusions. Questions about the exercise can be found at the end of the chapter.

Most manuals are workbooks in which you are expected to record data, write up your analysis, record your conclusions, and answer questions.

After answering the questions, use the SQ3R plus P to once again study the content of the laboratory manual. See chapter 7, "Time Management and Study Sessions" for other clues to studying the lab content.

Review

1. The lecture and lab content is the guide to what information you should study. If you do not have complete notes, then you will have difficulty knowing what to study in the textbook.
2. Your textbook will contain information and examples to reinforce and supplement the class content.
3. If information appears in lecture or lab notes and in your text or manual, then that is the important information to learn.
4. The assignments will help identify what you should study.
5. Before you use your text, you should formulate learning objectives. Use these to guide the use of the textbook.
6. You can learn from the text through the use of SQ3R.
7. SQ3R plus P means *Survey, Question, Read, Recite, Review* plus *Practice.*

Q U E S T I O N S

- What do the symbols in figure 10.1 represent?

- Record your answers in the added content column. Be sure to copy the symbol, then tell what that symbol means. Examples are given. It is always a good idea to incorporate the words or symbols of the question into your answer.

- The correct answers can be found in the appendix.

- There is no reason you should know all these symbols, but it would be interesting to test yourself.

A D D E D   C O N T E N T

Cu .... copper or cumulus cloud

ml .... milliliter

C H A P T E R

10

Words and Symbols

Objectives

1. To realize that specialized words and symbols are used to communicate scientific information.
2. To realize that you must be able to learn and use scientific words and symbols.

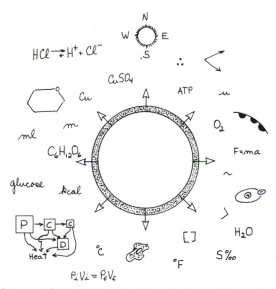

Figure 10.1 Examples of symbols used in science.

- Why is it important to learn the meaning of terms and symbols?

- Why are there so many new words to learn?

- What is the difference between knowing the meaning of a word and knowing how to use the word?

- Explain two ways suggested in this guide to learn new vocabulary words.

- How do you plan to learn the new words and symbols?

- Why are symbols used in science?

- How can a study group help you to learn new terminology?

- How can you identify what words and symbols you must learn?

- Explain the difference between knowing an equation and using an equation.

- Explain the terms *memorize* and *problem-solve*.

Words and Symbols

Through the years, scientists have investigated many different parts of the universe. Whether the science is astronomy, biology, chemistry, ecology, meteorology, or physics, each has developed a body of knowledge unique to its own area. New words and symbols have been formed to record and communicate this newfound knowledge. Units identify measurements of distance, time, energy, and mass. When different measurements are related, units of rate result. Meters per second, heartbeats per minute, moles per liter, and calories per gram are examples of rates.

As you study science, you must learn the words, symbols, measures, expressions of rates, and whatever else is used to communicate scientific information. If your instructor uses it, then you must learn it. Each word or symbol represents some specific thing. Remember, symbols are agreed-upon abbreviations to speed up the communication of information. New words are generally introduced in rapid fire (without mercy, it might seem). You are expected to be able to use a new term almost immediately. This is possible only if you repeatedly preview and review and practice the terminology outside class. Descriptive courses like anatomy might introduce forty to sixty new terms in one lesson. Problem-solving courses introduce fewer terms but apply basic concepts and relationships (rates and ratios) to different conditions or problems. The symbols in equations and the equations are of prime importance.

If you memorize the meaning of the words and symbols, then you will have the basic tools to communicate but not necessarily an understanding of the subject. The next challenge is to integrate these words and symbols into concepts, as expressed in complete sentences, paragraphs, solutions to problems, or complete thoughts. For instance, in the study of genetics, capital and lowercase letters are used and arranged in certain, specific ways (figure 10.2).

What does all this mean? How do the symbols relate to reproduction and the mechanisms of inheritance? (See appendix 1 for the answer.) The comprehension of the terms and symbols will enable you to answer questions, solve problems, and communicate in science.

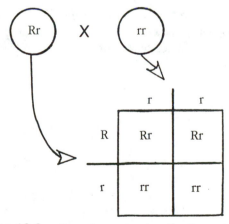

Figure 10.2   Genetic cross using the Punnet square.

Q U E S T I O N S

1. What kind of weather does Boston have?
2. What is the approximate temperature in Washington, Chicago, and Seattle?
3. What kind of weather front is off the east coast?
4. There are irregular bands that run from west to east. What do these bands indicate?
5. What was the temperature in your area on the day depicted on the map?
6. Where did snow fall in the country?

A D D E D   C O N T E N T

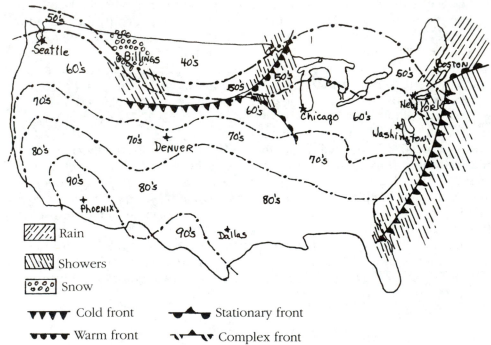

Figure 10.3   A weather map. Examples of symbols used in the science of Meteorology (the study of weather).

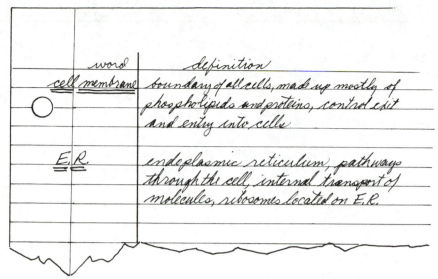

Figure 10.4   Example of a vocabulary-definition list.

Part of your study sessions should be devoted to the systematic learning of words and symbols. Decide on how you will do this and then do it. Two ways of learning keywords (test words) and symbols are to maintain a *running vocabulary-definition list* or to establish a *flash card system. The words to be learned are those recorded in the keywords column of your class notes.*

Vocabulary-Definition List

The word or symbol should be recorded in a column on the left-hand side of a page; the definitions are in a column on the right-hand side. You can refer to the glossary or your notes before expressing the meaning *in your own words.* Review the list frequently. To test your knowledge, cover either the word or the meaning. Place a check mark next to the words you understand (figure 10.4).

Flash Cards

Write the word on one side of the card. Don't put more than one word on a card. (Use paper cut into four pieces rather than buying cards. It's cheaper and just as good.) On the other side, record the meaning and related information. For instance, a cell membrane card might include the above information and reference to osmosis, active transport, endocytosis, and exocytosis. Carry your stack of cards with you. Whenever you have a few free moments, review the cards. Begin to separate the stack into "known" and "not known" piles. Continue to work on the "not known" pile but review the "known" stack every once in a while (figure 10.5).

- Do you think you could use flash cards to help you memorize information in the science course you are taking?

- Explain when and how you should make up flash cards.

- How do you plan to learn how to comprehend and use the words and symbols you memorize?

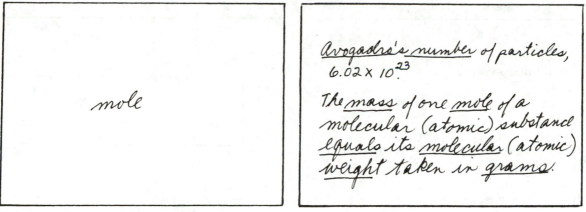

mole

Avogadro's number of particles, 6.02 × 10²³

The mass of one mole of a molecular (atomic) substance equals its molecular (atomic) weight taken in grams.

Figure 10.5   An example of a flash card.

Take special note of the way your instructor uses symbols. Arrows, letters, figures of different shapes and shading, and specialized symbols are all important in the communication of information. Frequently these are combined to form figures that describe a process of some sort. Below are examples of different symbols used in science.

1. Arrows ⟶

In physics this represents the *direction* in which a force is applied; in chemistry it represents the *direction* of a chemical reaction; in biology it represents the *change* of an organism as it develops, such as from an egg to a tadpole to a frog; in ecology it *shows the flow* of energy or resources; in molecular biology it means that DNA *synthesizes* mRNA.

2. Letters and numbers

m	Represents mass or a meter.
Cu	The chemical symbol for copper or the symbol for a cumulus cloud.
U_{235}	The symbol for uranium and its mass number. Note that the number is a subscript.
0° C	The degrees Celsius at which water freezes.
ATP	A molecule of adenosine triphosphate.
$C_6H_{12}O_6$	One molecule of glucose made up of six atoms of carbon, twelve of hydrogen, and six of oxygen.

3. Examples of specialized symbols

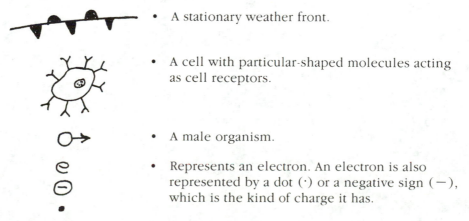

- A stationary weather front.

- A cell with particular-shaped molecules acting as cell receptors.

- A male organism.

- Represents an electron. An electron is also represented by a dot (·) or a negative sign (−), which is the kind of charge it has.

| QUESTIONS | ADDED CONTENT |

- What do the different symbols (different-sized lines and numbers) represent on these rulers?

- Which represents measurements in the metric system?

- Which is the system we typically use around our homes in the United States?

- Which system is used in science labs?

- What does each of these thermometers measure?

- Which probably measures degrees Celsius (centigrade), and which measures degrees Fahrenheit?

- What number of degrees does each unit of space represent on the thermometers?

- At what temperature does water freeze?

- What is normal body temperature in degrees centigrade?

- What does *centi-* mean?

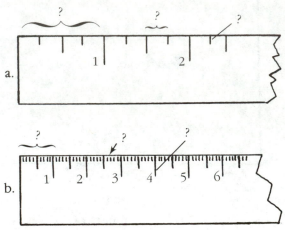

Figure 10.6   These are parts of tools to measure length. What do the symbols mean?

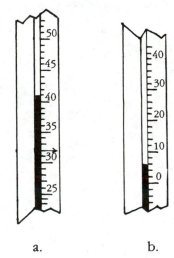

a.   b.

Figure 10.7   Diagrams of parts of two different types of thermometers.

Review

1. Each science has developed a specialized vocabulary.
2. Symbols are used to express the content and concepts of the science in shortened form.
3. It is important to learn the terminology and the symbols of science.
4. It is important that you be able to use the vocabulary of science to express complete and logical thoughts.
5. Recording of keywords in notes and the use of vocabulary-definition lists or flash cards help you identify and learn the important terms and symbols.
6. Symbols take many different forms and are often incorporated into figures that describe the content of science.

- What does each box and circle represent in figure 11.1a?

- What do the arrows represent?

- What kind of ecosystem is pictured in the lower left part of the small illustration?

- What is an ecosystem?

- Give examples of species of organisms that have fairly stable populations.

- What does figure 11.1a indicate is happening to the human population?

- What is the difference in how matter and raw materials are managed by the two different ecosystems?

- What kind of organism converts solar energy to chemical energy? (The answer to this question is not found directly in the figure.)

- What kind of energy is used by the human population?

- What does part b of figure 11.1 diagram?

- What percentage of the population are females who are twenty to twenty-four years old?

- What percent of the human population are males who are seventy to seventy-four years old?

- Does figure 11.1b depict the actual population of the United States?

11

Figures and How To Use Them

Objective

1. To learn what are the different types of figures used in lectures, textbooks, and laboratory manuals.
2. To learn how to analyze figures in texts.

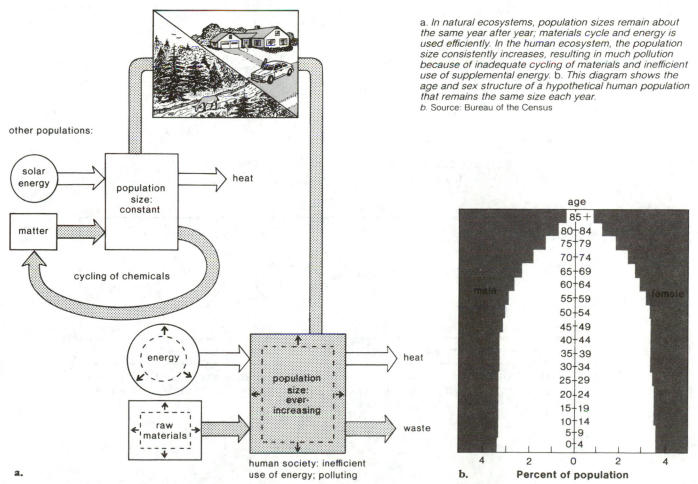

a. *In natural ecosystems, population sizes remain about the same year after year; materials cycle and energy is used efficiently. In the human ecosystem, the population size consistently increases, resulting in much pollution because of inadequate cycling of materials and inefficient use of supplemental energy.* b. *This diagram shows the age and sex structure of a hypothetical human population that remains the same size each year.*
b. Source: Bureau of the Census

Figure 11.1 Format of figures in science textbooks.

Q U E S T I O N S   A D D E D   C O N T E N T

What do parts a and b of figure 11.2 relate?

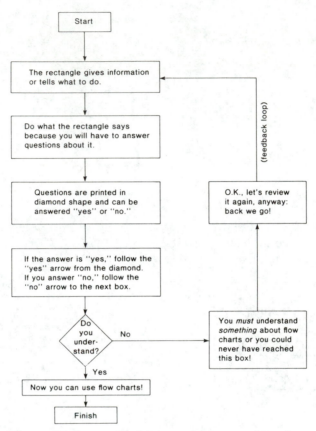

Figure 11.2a   A flow chart on how to use flow charts. Follow from start to finish.

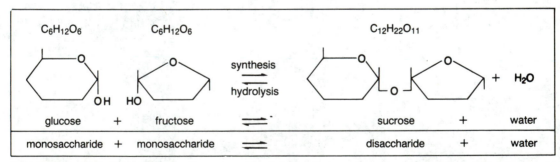

Figure 11.2b   Notice in this diagram each of the organic molecules is represented in four ways, i.e. $C_6H_{12}O_6$, the shortened structural diagram, glucose and monosaccharide. There are at least 15 bits of information in this figure. How many do you see?

A picture is worth a thousand words!

Analyze a figure, then use SQ3R to read the thousand words. Figures will help you visualize the instructor's spoken word and the textbook's written word. *Understanding figures is an important study skill,* as is the ability to construct diagrams.

Make every effort to understand the figures used or referred to by your instructor. All too often, students ignore or skip over these valuable study aids because they look too "difficult." Part of the *survey* of SQ3R should be the analysis of figures in your textbook or notes. Part of the *3 Rs* of SQ3R should be deciding the relationships between the content of important figures and the content of class notes. *Important figures* are the figures used in the lecture and lab and figures in the textbook and lab manual. These figures will symbolize the important information covered in their course.

Types of Figures

The types of figures used include maps, photographs, illustrations, diagrams, charts, tables, and graphs. The discussion in any textbook refers to various figures. Each figure is *numbered* and has a *caption*. The caption describes the main purpose of the figure. The third part of the figure is the *body or content* of the figure. Close examination of the symbols used in many figures often yields more information than is included in the caption.

Maps

Maps are used to relate scientific information to continents, nations, or some other geographic area. Weather, ecological, or geological features are represented on maps. See the weather map on page 74 as an example of this type of figure.

THE FAR SIDE By GARY LARSON

Figure 11.3 An example of a map. Which areas (belts) are not identified?
(The Far Side. Copyright 1991 Universal Press Syndicate. Reprinted with permission.)

QUESTIONS

- Where is Devil's Tower?

- What is Devil's Tower?

- Explain how Devil's Tower was formed.

- What do these bar graphs relate?

ADDED CONTENT

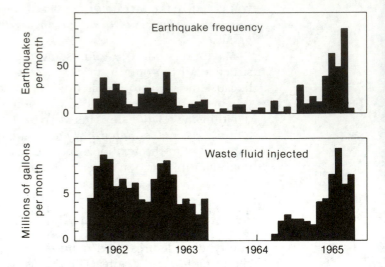

Correlation between waste disposal at the Rocky Mountain Arsenal (lower diagram) and frequency of earthquakes in the Denver area. From David Evans, "Man-made Earthquakes in Denver," in *Geotimes*, Vol. 10. Reprinted by permission of the author.

Figure 11.4   Bar graphs can summarize quantitative information in a small space.

- What rates are indicated on the y axis (ordinate or vertical line)?

- What unit of measurement is indicated on the x axis (abscissa or horizontal line)?

- What was the source of the waste fluid injected into the ground?

- What information does figure 11.5 illustrate?

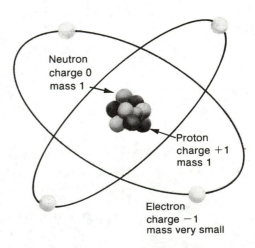

Basic atomic structure: protons and any neutrons in the nucleus, electrons circling outside it.

Figure 11.5   A sample diagram representing something we can't see but we know exists.

A volcanic neck or pipe—all that remains of a long-eroded volcano. (a) Schematic diagram: Note cylindrical shape and discordant character. (b) An example exposed at the surface by erosion: Devil's Tower, Montana.

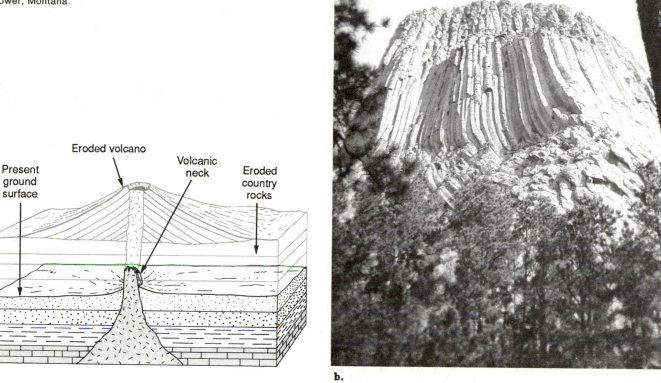

Figure 11.6 This is an example of an illustration (*a*) and a photograph (*b*).

Photographs and Illustrations

Photographs are used to provide views of things described in the text or to perk up your interest. Illustrations are artistic renditions of something the author thinks will enrich his or her description. Illustrations and photographs might be placed next to each other for easy comparison. This format is common in geology and environmental science textbooks (figure 11.6).

Diagram

Diagrams are simplified illustrations. Components are seldom to scale. Diagrams could be realistic outlines of objects or they could be generalized impressions of objects. Atoms, electric circuits, telescopes, microscopes, cell parts, organ systems, weather systems, and strata of rock are all examples of things that might be diagrammed. Frequently diagrams express relationships between things. Examples include flow, organization, process, tree, and comparison diagrams. A life cycle, rock cycle, or nutrient cycle might be represented in a diagram. The processes of nuclear fuel mining and refining are represented, as is the flow of energy through an ecosystem (see figure 11.1). Arrows, different-shaped symbols, words, and simple illustrations might all be included in a diagram.

• How well can you analyze a figure?
Practice on the example below.

a.

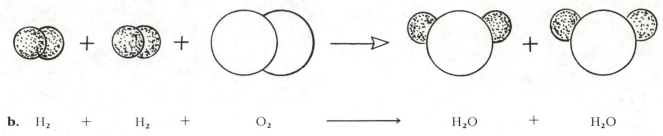

b. H_2 + H_2 + O_2 $\longrightarrow$ H_2O + H_2O

c. $2 H_2 + O_2$ $\longrightarrow$ $2 H_2O$

Figure 11.7   The meaning of a balanced chemical equation is when the number of atoms (by themselves or in molecules) that appear to the left is equal to the number of atoms on the right of a chemical equation. The atoms are rearranged in a chemical reaction.

Before you answer the questions on page 8, be sure you have spent time practicing the steps of how to analyze a figure. Make up your own questions, and then compare them to those asked about the figure above. Good learning skills include good questioning skills and practice of the skill.

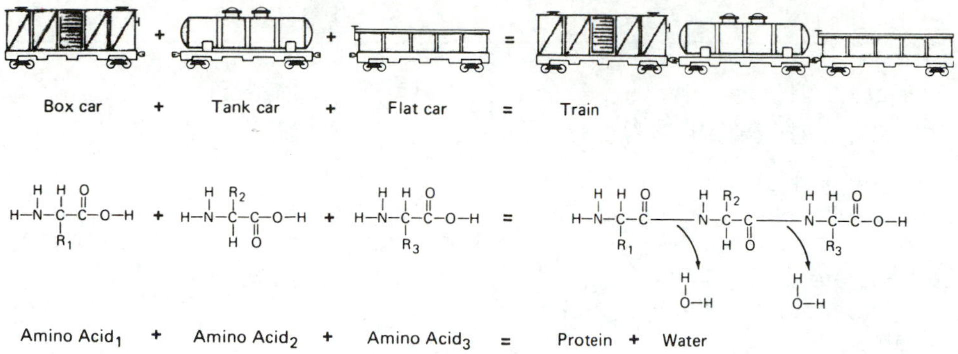

Box car + Tank car + Flat car = Train

Amino Acid$_1$ + Amino Acid$_2$ + Amino Acid$_3$ = Protein + Water

Figure 11.8 Words, symbols and simple diagrams represent defferent ways to express what is written in this caption.

Tables

Tables organize information into rows and columns. Tables are used to compare data or to summarize characteristics of lists of different components. The ability to construct an information table is a valuable study skill. Learn to reorganize parts of your class notes into tables. Examples can be found on pages 46 and 118.

Graphs

Graphs usually compare two factors in a general or quantitative fashion. A quick analysis of a graph gives you very specific information or enables you to see relationships between two or more factors. Forms of graphs include bar, pie, line, and pictographs. Most graphs have units clearly labeled in the caption, in the legend, or most frequently on the graph itself.

Line graphs are of particular importance in biology, chemistry, and physics. If you are constructing a graph, remember to

1. Title the graph and label each axis. The y axis is the vertical axis, and the x axis is the horizontal axis.
2. Clearly identify magnitude and types of units on each axis.

Questions about figure 11.7.

1. Did you underline the keywords in the caption?
2. What is the meaning of a balanced equation?
3. What symbol separates the equation into left and right sides?
4. In a balanced equation, what is the relationship of the numbers of atoms on the left and the right?
5. What do the symbols H, H_2, O, O_2, and H_2O represent?
6. How does the illustrator represent H_2, O_2, and H_2O?
7. Why are the above atoms and molecules drawn and shaded the way they are?
8. What do the (+) and the ⟶ represent?
9. Why are there three representations (a,b, and c) in this figure?
10. What does the 2 in front of H_2 and H_2O represent in part c of figure 11.7?
11. Why is there no 2 in front of O_2?
12. How are the atoms rearranged in this chemical reaction?
13. Why are the hydrogen atoms in the water molecule placed where they are rather than on exactly opposite sides of the oxygen atom (Mickey Mouse vs. Alfred E. Newman)?
14. Summarize what numbers and types of reactants will form what numbers and type of product in this balanced chemical reaction.
15. Make up a list of keywords that you would predict could be used on a test about this figure.

Answers to these questions can be found in the appendix. As you check your answer, note the *content* of the *answer* and the style or method of writing an answer.

3. When drawing the line, generalize it; don't connect one data point to another. Use a French curve if the lines are curved.
4. Extend the line slightly beyond the first and last data points.
5. Use the whole sheet of graph paper, not just a corner. This means that the units have to be scaled out to the ends of each axis you draw.
6. Surround the data points with a circle, square, or triangle if several data lines are drawn on one graph.
7. Work neatly.
8. If you have trouble with the technique of making a graph, seek help from the instructor or science tutor.

Figure analysis is a study skill you must develop and use! Practice analyzing figures. After reading about how to analyze a figure below, practice explaining figures to a friend.

Remember—a picture is worth a thousand words.

Figure 11.9 How does this cartoon relate to the information in figure 11.7? (By permission of Johnny Hart and Creators Syndicate, Inc.)

Figure Analysis

When you analyze a figure, have a pencil in hand to help you "get into" the details of the figure. Follow these steps:

1. Survey the figure to get a general feel for the figure. Ask: *"What is this figure about?"*
2. Read the caption. Underline the keywords.
3. Recite or visualize the meaning of the keywords. If you have trouble with this, refer to the text or the glossary to clarify the meanings of the terms.
4. Relate the caption to the figure by identifying the content of the caption in the figure. Identify all of the labeled symbols or components. Use your pencil point to locate them. Point to *everything in the figure.*

Q U E S T I O N S	A D D E D C O N T E N T

- Name three parts of a figure in a textbook.

- Why are figures used so frequently in science books?

- List the different kinds of figures that can be in a science text.

- Which of the above types of figures can you find in your text?

- What is the difference between a pie graph and a line graph?

- Summarize the eight steps that are recommended to help you analyze a figure.

- Does this system of analysis seem helpful? Why or why not?

- Did you check the appendix after thinking about the genetic symbols in the previous chapter? If you did, . . . great! If you didn't, then you really have not developed the questioning attitude. You were trying to cut corners.

5. Now use your pencil to point out all other parts of the figure not identified. Try to decide what they represent.
6. After doing this, review the content of the figure.
7. Ask yourself questions about the figure. Use the figure to answer the questions you generated.
8. Finally, relate the figure's content to the material covered in lecture, in lab, or in your notes.

Review

1. Figures are important aids to communicate scientific information.
2. You should develop and practice the skills that will enable you to analyze and comprehend different types of figures.
3. Figures include a number, a caption, and a body of information.
4. When you analyze a figure you should:
   a. survey the figure;
   b. read the caption;
   c. recite or recall the meaning to keywords;
   d. relate the caption to the figure;
   e. identify and comprehend all the labeled and unlabeled parts of the figure;
   f. review what the figure explains;
   g. ask yourself questions about the figure and answer them;
   h. relate the figure to the lecture or laboratory content.

| QUESTIONS | ADDED CONTENT |

- What do authors of science textbooks say about the characteristics of students who are good problem-solvers?

Bodner and Pardue, who wrote *Chemistry, An Experimental Science* (Wiley, 1989), list the following characteristics in their new textbook.

Good problem-solvers:

1. Believe they can solve almost any problem if they work long enough.
2. Read carefully and reread a problem until they understand what information is given and what they are asked to solve for.
3. Break problems into small steps, which they solve one at a time.
4. Organize their work so that they don't lose sight of what they've accomplished, and can follow the steps they've taken so far.
5. Check their work, not only at the end of the problem, but at various points along the way.
6. Build models or representations of the problems that can take the form of a list of relevant information, a picture, or a concrete example.
7. Try to solve a simpler, related problem when faced with a problem they can't solve.
8. Guess and test; they try out several approaches to a problem until they are successful.

12

Problem-Solving

Objectives

1. To learn a systematic approach to solving problems presented in a science course.
2. To recognize that some problems require verbal solutions and others require mathematical solutions.

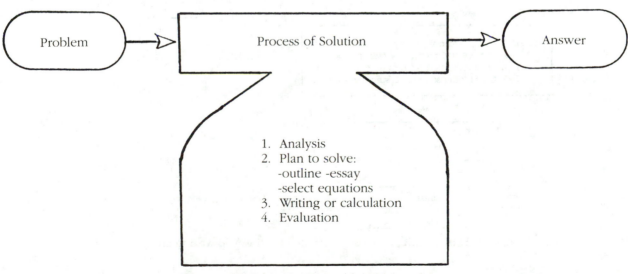

Figure 12.1   Components of problem-solving.

Problem-solving is a skill that can be developed. If you develop a systematic approach to problem-solving and practice that approach, you will improve your problem-solving skills. *Practice*—which means involvement, time and effort—does help!

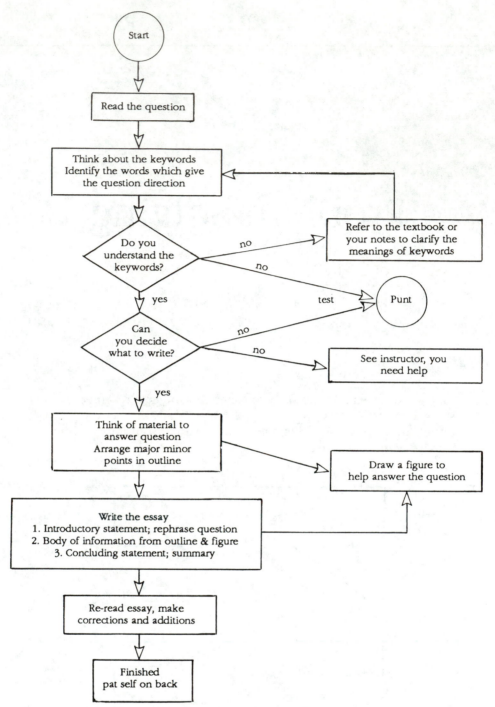

Figure 12.2   This demonstrates a pathway to writing an essay.

Process of Problem-solving

As you read a question, it is important to comprehend the information in the question and to identify what *problem* you are being asked to resolve. Problem-solving involves

1. Analysis of the problem;
2. Planning a solution;
3. Answering the problem according to a plan;
4. Evaluation of the solution.

Answering any question involves a problem-solving procedure. Short-answer questions such as multiple choice and true-false are problems posed about rather discrete bits of information, relationships, or concepts. Hints to answering these types of questions are given in the chapter on test-taking. Nevertheless, what is said in this chapter applies to solving short-answer types of questions also.

Essay and Math Questions

An *essay question* (figure 12.2) is a problem that can be answered (solved) verbally. The *construction* of an *outline* is the key to writing a good essay answer. Other questions pose problems that require a *mathematical solution* (figure 12.3). In the math-oriented problems, the selection of an *equation* or a properly *sequenced series of equations* is the key to calculating the answer. Instructors of most introductory science courses could ask you to solve essay problems. Courses in chemistry and physics are quantitative. They are math based; solving math-based problems is a major part of the course. You will apply principles as expressed by equations to problems.

1. Analysis
    As you begin the *analysis* you should:

    *   Read the question carefully. Think about and define the keywords and their relationships. Identify any words or phrases that give a clue to a solution, such as "sum of," "equal to," "products of," "constant temperature," "compare," "differentiate." (See figure 14.3.)
    *   Identify the problem to be answered, and define what principles or concepts the problem involves.
    *   If it is a math problem,
        a. Write down what is given (known) in the question and what is wanted (to be found).
        b. Draw a simple diagram and label of what is given and wanted.
        c. Identify whether part of the solution requires the conversion of units. Do this before you begin to solve the problem.
2. Planning

In the planning stage, think about the relationships and information that connect the given and wanted information. For essay questions, you should construct an *outline* of information that answers the question. Major topics of the answer should be listed. Each major topic should be supported by details or examples. *Do not assume that your instructor will think you know the answer.*

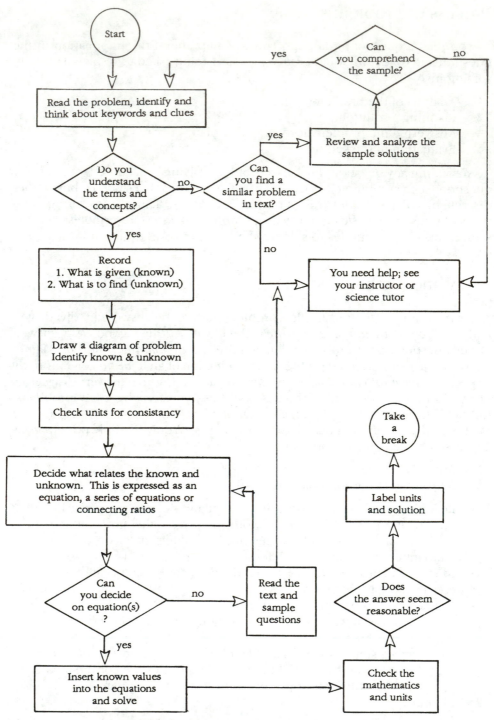

Figure 12.3   This is a pathway to solving problems involving mathematics.

For math questions, you will have to decide on which *equations* will relate the known and unknown factors. If more than one equation is required, then be sure to plan the sequence of how to apply the equations. If you can integrate the equations, then do so before you calculate the answer. Remember, the equation actually describes a system or process in the universe and relates the different parameters of the described system.

3. Answering—Writing and calculating

After all this is accomplished (the analysis, the planning), the bulk of your work is done. It is now a matter of incorporating what you have planned into either an essay or a calculation.

The essay answer should be written in three parts.

1. Begin the essay with a thesis sentence, sentences, or paragraph. This beginning should restate the problem and indicate how you will answer the question. A question from a geology book states: "Describe two sources of heat causing metamorphism." A beginning of an essay might read: "Two sources of heat causing metamorphism are . . ."

2. The body of the essay contains the supporting information. The detail of the outline should make this easy to write. Don't add a lot of bull to embellish the essay. *Use the vocabulary of science* to indicate what you have learned. If a diagram is appropriate, include one, but bear in mind that a diagram is not an acceptable answer to an essay question. However, it will enrich an answer to the question. Leave a few lines between paragraphs just in case you might want to add something after you have reread the answer.

3. The concluding sentence, sentences, or paragraph should summarize how you have answered the problem. "Thus the heat to form metamorphic rock can come from the geothermal gradient and by plutonic activity . . ."

QUESTIONS   ADDED CONTENT

Physics problem

A metal ball is dropped from the top of the Empire State Building. The ball takes 8.9 seconds to reach the sidewalk. Determine a) the velocity of the ball when it hit the sidewalk, b) the average velocity of the ball, and c) the distance the ball fell from where it was dropped to the sidewalk.

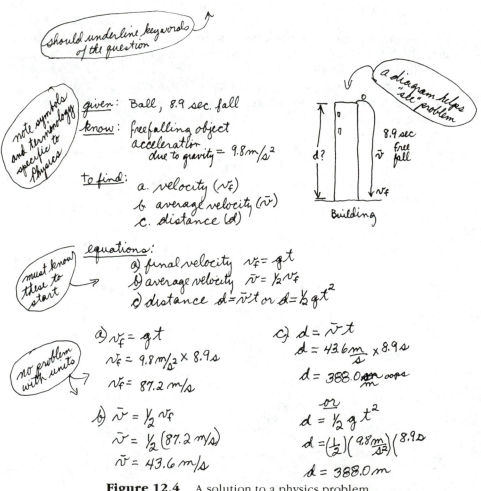

Figure 12.4   A solution to a physics problem.

Math-based Problems

If the problem is math based, (figures 12.3, 12.4) then insert the known values into the equation and do the calculations to solve for the unknowns. Before proceeding with the calculations, however, you should once again check the units of the known variables. The units of the given information should fit into the equation. If mass is called for, then make sure you don't try to insert a unit for rate of some sort. Realize that one unknown in a problem will require one equation in the solution. Two unknowns will require two equations. When the calculation is complete, be sure to label the answer, designating units.

The most common errors in mathematical problems involve

- wrong units or improper unit conversion;
- use of the wrong equation;
- solving for the wrong variable;
- mistake in the equation or calculation;
- selection of improper information from a complex problem;
- lack of ability to relate a sequence of mathematic expressions to a problem.

4.   Evaluate

Finally, *evaluate* your answer. The answer should make sense. Reread the essay. Compare the content to the outline; make sure nothing was omitted. Check your calculations for mathematical errors. Decimal marks have a tendency to be misplaced. Make one final check to see that what was wanted in the analysis was given in the solution. *Be sure that you answered the question that was asked,* not one that you made up.

QUESTIONS	ADDED CONTENT

What are examples of good and poor answers to questions?

Explain why the bottom of a shower curtain moves inward when a person takes a hot shower.

The molecules are different. The steam pulls the curtain in because the cold air pushes on the hot air.

When hot air rises, it flows across the room, then is pushed down as it cools, then pushes against the shower curtain, heats up, and rises again. This is called a convection current.

The hot water heats the air in the shower. Air (gases) which become warm decrease in density. This less dense warmer air rises to the top of the shower and passes out into the room. A convection current of air is started. Cooler, denser air on the floor and base of the shower begins to move into the shower to replace the air which has risen to the top. As this cool dense air moves it pushes the shower curtain inward. The warm air cools, this air moves downward as it cools and becomes denser. A convection current of air is established.

Figure 12.5   These answers are arranged in order of quality, poor to good. Note how the best answer uses scientific terms like density, convection current, and gases.

Four words are listed. One word in the group can be considered different. Circle the different word and explain how it is different. Explain what links the other three.

a) carbon, water, hydrogen, oxygen

water is a liquid the rest are gases

a) carbon, (water), hydrogen, oxygen
Each one carbon, hydrogen, and oxygen are distinct gases. Water (H₂O) is a combination of the three

a) carbon, (water), hydrogen, oxygen
Water is a compound composed of two elements. The other three are elements

Figure 12.6   Can you pick out the best answer? Is carbon a gas? Are oxygen and hydrogen always gases?

Review

1. Read the problem . . . carefully! Then read it again.
2. Think about and identify the content of the problem.
3. Identify what is given and what is wanted.
4. If units are given in the question, check to see whether they are compatible, part of the same system of measurement.
5. Plan a solution for an essay question by constructing an outline.
6. Plan a solution for a math-based question by selecting equations that relate the known and unknown parts of the problem.
7. Write the essay. Calculate the answer.
8. Check work by rereading or double-checking the mathematics.

QUESTIONS	ADDED CONTENT

- Go back to the study skills checklist at the beginning of the book. Evaluate how well you perform these study skills on a scale of 1 (poor) to 10 (good).

- Are you trying to improve those skills rated poorly? How?

- Diagram an information map representing what you do to learn.

- Diagram an information map representing what the above *added content* discussion suggests should be the ingredients of activities to help you learn.

- Compare these two information maps.

- Which involves a greater variety of study methods?

The "study skills inventory" indicates that there are many different ways to study the same material. *Study* is a series of *organized activities* directed to learning. You will find that if you engage in previewing, listening, note-taking, reading, reciting, reviewing, group study, question generation, problem-solving, self-testing, recalling and discussion with peers, tutors, and instructors then you will learn and comprehend scientific information.

Information map. Your method of study.

Ingredients to study and learning as suggested by the above added content discussion.

13

Types of Tests—
Preparation

Objectives

1. To learn about the different kinds of tests generally given in science courses.
2. To review what should be done to prepare for tests.

Types of Tests

The information table and map in figure 13.1 pictures information about different kinds of tests. Note that the table and map give the same information, but the information is organized in a different way. Which format is easier to analyze? Remember, tables and maps are ways to reorganize your class notes into summaries of information. These types of summaries would allow you to efficiently and effectively prepare for a test. It's easy to review material in tables, maps, charts, and outlines.

Tests can be either short-answer or long-answer. Short-answer questions can be constructed to test recall about specific information or general concepts. Word recognition is important for these types of questions. In addition, this format, when properly constructed, can test your understanding of complex relationships. When done correctly, the questions are long and loaded with scientific terminology. Good reading skills are important to successfully answer these questions.

Short Answers

Identification and fill-in questions require you to recall words or phrases and to record these answers on the answer sheet. These questions are used most frequently on laboratory tests. Biology and geology identification tests are often

- Review the suggested format for taking class notes in chapter 8.

- Did you evaluate your class notes by filling in the notebook checklist?

- How does the format of your notes automatically help you to prepare for a test. Or are your notes a jumble?

- Do you keep a keyword list in your notes?

- Are figures in your notes understandable? Could you explain them to someone if you needed to?

- Did you take notes on the instructor's discussion as well as on what he wrote on the board?

- Did you write a brief summary of each day's lecture in your notes?

- Do you compose a summary at the end of the week for each week's work?

- Have you become part of a study group?

- Have you recorded questions that your instructors asked in class?

- Have you generated questions about the content of the lecture and textbook?

- Have you been studying science six to ten hours per week?

- Have you practiced problem-solving?

Test types		Format	General use
Short answer	written	Identification........ Fill in....................	Laboratory Lab & lecture
	choices	True-false Multiple choice Matching	Lecture or Laboratory
Long answer		Sentence or two Paragraph Essay	Lecture or Laboratory
Mathematical		Problem solving	Lecture or Lab

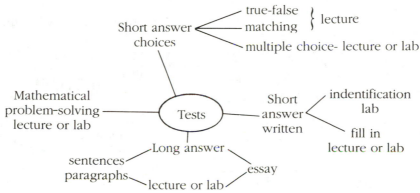

Figure 13.1 Information map and table—types of tests.

spread around a room. You will move from question to question. After reading the question, you will have about a minute to answer it before moving on to the next question.

True-false and multiple-choice questions allow you to choose from answers given on the test. In a *true-false* format, you have to select whether a statement is one of two choices: true or false. *Multiple-choice tests* will require you to select from four or five possible answers. Variations of this format could often have answers like "all of the above" or "none of the above." Sometimes variations like "answers 'a' and 'c' are correct" or "b and d" are on tests when a combination of choices is correct. Read these types of questions with particular care. Multiple-choice questions can be considered to be a series of true-false questions packaged into a single question. *Read all choices* with care to choose the *best answer.*

Matching questions require the matching of terms, phrases, or sentences in one column to terms or phrases in another column. If you know all the keywords (test words), you should do well on this type of test.

Long Answers

Long-answer questions require you to write a sentence or more. Instructors might ask you to define a term like *osmosis, pressure,* or *vector.* On the other hand, you might be asked to briefly explain "under what conditions water will be lost from a cell" or "the relationship between pressure and volume of a gas when temperature is held constant." All of the above require relatively short written answers.

Q U E S T I O N S	A D D E D C O N T E N T

- How should you go about improving your study skills and your preparation for tests?

1. Try to keep things simple. Don't defeat yourself by studying everything. Study the topics the instructor covered.

2. Establish specific objectives. Answer the *how, when, where, what,* and *why* questions you have generated.

3. Break big tasks down into smaller ones.

4. Practice solving problems!

5. Study in a study group.

6. Study frequently; repeat learning the material in different formats . . . notes, figures, summaries, charts, maps. Use tutors.

7. Realize when you study you're doing something good; it is good for your academic health.

8. Assume the responsibility to study. Remember, it is all yours. *Excuses don't count.*

9. Don't delay, schedule six to ten hours each week to study science.

10. Practice the study skills until they become second nature to you. Internalize the study skills.

11. Motivate yourself to learn the subject material.

12. Realize it is satisfying to become competent in an academic area.

Essay questions (see pages 94 and 118) require you to write about concepts supported by specific information or examples. The answers to these questions might require several pages of written work. If the instructor indicates that essays will be on the test, then you should try to get him to give examples of the types of questions he will ask. It doesn't hurt to ask. The reply might help you to prepare for the test.

Preparation

If you have established a cycle of weekly study, if you define study objectives and accomplished them, if you have organized weekly summaries, if you have studied with classmates, and if you have generated your own questions, then studying for a test will be relatively easy. Cramming or some other last-minute study system isn't worth talking about. Remember—effective learning involves frequent reviews to comprehend and remember the information. Frequent reviews allow you to relearn forgotten information. The *final review* before a test just sharpens what you have already learned.

> Begin to study for a test two or three days
> before the test.

You should begin studying for a test by finding out as much about the test as possible.

- What kinds of questions will be asked?
- How many questions will be asked?
- How long will you have to finish the test?
- Do you need to bring pens, pencils, or special answer sheets?
- What are the topics to be covered on the test?
- Will you be able to keep the questions for review purposes?

The answers to these questions will help you study for and take the test. For instance, it makes a big difference if you know whether you are going to have forty or sixty short-answer questions to finish in fifty minutes. Forty questions will mean you'll have over a minute per question, whereas sixty questions will give you less than a minute per question.

If the test is to be short answers, then you must not only study the overall concepts and interconnecting principles, but also the details discussed in lecture. The majority of short-answer questions will be about specific information, concise statements of concepts, and application of principles to specific examples.

If essay questions are to be answered, then the overall approach to the course content must be reviewed and crystallized. You should be able to apply concepts

If you were taking a chemistry course, then acids and bases would be one of the topics you would study. Questions you generate could serve as your study objectives. *Constructing numerous questions also breaks the broad topic into smaller, more manageable units to study.*

These questions have been generated from *headings* and *subheadings* in a *chemistry textbook*. Others could have been generated from the bold-printed words. Still other questions could be fashioned from the sample problems given in the content of the chapter.

Following are examples of questions you would have to answer during your preparation for a test on acids and bases:

1. What are the characteristics of an acid and a base?
2. How do acids and bases differ?
3. Give the name and molecular formula for three acids and three bases.
4. When the above acids are dissolved in water and dissociate (ionize), what ions are formed?
5. When the named bases are dissolved in water and dissociate (ionize), what ions are formed?
6. Differentiate between the Arrhenius and Bronsted-Lowry definitions of acids and bases.
7. Give examples of strong acids and bases.
8. What is a polyprotic acid?
9. What is neutralization?
10. Give examples of three neutralizations reactions, and make sure they are balanced.
11. Explain the pH scale. Give the average pH of some common liquids.
12. Name three indicators of acids and bases. Describe the colors they exhibit at different pHs.
13. How many milliliters of a 2M HC1 solution would just neutralize 25 milliliters of a 1M NaOH solution?

and principles and support them with examples. If essay questions are to be asked, then you should try to predict what they might be. Your instructor might give a few hints before the test. Listen for clues. If an instructor spends a week enthusiastically discussing a topic, you should be able to write an essay on what has been discussed. Obviously if an instructor gives you sample questions, then you should prepare answers for all the questions.

If mathematical problem-solving is to be on the test, you must *practice, practice, and practice* solving problems as you prepare for the test. Be sure you can solve examples of each type of problem discussed in class. Remember, these types of tests apply the principles studied in class and in the textbook.

To prepare for a test, you should

- Study the weekly summaries you have organized.
- Review all of your notes and identify areas you feel uncertain about. Concentrate on these unfamiliar areas.
- Survey pertinent parts of the book (those areas covered in lecture).
- Practice solving problems.
- Continue to work with flash cards or vocabulary-definition lists.
- Test yourself with questions from the questions column of your notes and the questions that appear in the textbook.
- If sample tests are available, then take these at least two days before the test. Study the content of the questions you got wrong. Don't just memorize the right answers.
- Meet with a study group at least two or three days before the test.
- Be sure to analyze and be able to explain all figures used by your instructor.

Day Before the Test

The day before the test, you should be able to sit down and visualize and recall the information presented to you week by week. After trying to recall this information, you should once again review the summaries, survey your notes, reproduce figures, skim the textbook, and review the questions. Take a break. Then skim your questions one last time and try to predict which questions will be on the test. Make up your own test and answer it.

Get a good night's sleep! Sleep on what you have learned.

QUESTIONS

What are some examples of test questions that appear on science tests? (Answers are in the appendix.)

Practice the beginning of test-taking by identifying the keywords in the sample questions below. Try to answer the question, but don't feel bad if you can't answer the questions. You enroll in a science course to learn new information and concepts, not just to repeat what you have learned in the past.

In the *true-false* questions circle the *T* or *F*, depending upon whether the statement is true or false. In the *multiple-choice* questions, circle the choice that best answers the question or completes the statement.

Oceanography

T or F

1. The foreshore is that portion of a beach that is never covered by water even at high tide.
2. Which of the following is in the middle layer of a stratified body of water?
   a. epilimnion   d. homothermous
   b. hypolimnion   e. lithosphere
   c. thermocline
3. In which areas of a body of water are conditions best for phytoplankton growth?
   a. hypolimnion   d. euphotic zone
   b. biota   e. nekton
   c. autotrophic

Meterology

T or F

4. A bar is a unit of pressure equal to 1,000 millibars or 29.53 inches of mercury.
5. Thunder is
   a. the clash between two colliding weather fronts.
   b. the side effects of the tremendous energy released by a discharge of lightning.
   c. caused by cumulonimbus clouds surging upward against a cold air mass.
   d. the results of the aftereffects of a tornado
   e. none of the above.
6. A warm air mass moving toward and lifting over a cold air mass is called a
   a. stationary front.   d. warm front.
   b. occluded front.   e. moving front.
   c. cold front.

Geology

T or F

7. In undisturbed layers of sedimentary rock, the oldest rock is on top, the youngest on the bottom.
8. Mechanical weathering is caused by all the following except
   a. abrasion.
   b. frost wedging.
   c. weak acid.
   d. frost heaving.
   e. pressure release.
9. Quartzite is the metamorphic product of
   a. quartz sandstone.
   b. granite.
   c. limestone.
   d. shale.
   e. rhyolite.

Biology

T or F

10. Plants are multicellular, eukaryotic, and autotrophic and have cell walls.
11. Hormones and pheromones are organic molecules that function to
    a. deliver some sort of message.
    b. provide the organism with energy.
    c. protect the organism from pathogens.
    d. act as molecular building blocks.
    e. serve as cell receptors.
12. The DNA code is based on a triplet of
    a. amino acids.
    b. phosphates.
    c. deoxyribose.
    d. ATPs.
    e. nucleotides.

Chemistry

T or F

13. In forming an ionic bond, each of the reacting atoms shares two or more electrons between the atoms.
14. Fill in the correct answer.
    What is the molarity of chloride ions in these solutions?
    a. 0.10 M $NaCl$ _____
    b. 0.10 M KCl _____
    c. 0.10 M $BaCl_2$ _____
    d. 0.10 M $AlCl_3$ _____

Review

1. Tests have questions that require short or long answers.
2. You should find out as much about the test as possible before the test. Do not be surprised by the format of the test.
3. To be able to answer essay questions, develop good outlining and writing skills.
4. Practice solving math-oriented questions.
5. Successful studying for tests depends upon establishing a cycle of weekly study.
6. Cramming is a slipshod method of learning.
7. Begin studying for a test two or three days before the test.
8. Study and review summary charts and maps, class notes, text, figures, and questions.
9. Review difficult areas first.
10. Learning depends upon frequent reviewing to enable you to relearn forgotten information.
11. Frequent study sessions spread the preparation out, thus making the study easier; it will seem like less work.
12. Studying frequently breaks big tasks into smaller ones.
13. Review the material the night before the test, but then give yourself time for a good night's rest.

Checklist for Taking a Short-Answer Test.

Check which of the following test-taking skills you use.

_____ Practice some form of relaxation before the test.

_____ Come to test on time but not early.

_____ Bring extra pencils and pens to test.

_____ Listen attentively to verbal additions and corrections.

_____ Make corrections on the test before starting to take the test.

_____ Read the directions.

_____ Survey the entire test before taking the test.

_____ Use a pencil as you read the test to guide attention to the question and the keywords in the question.

_____ Underline keywords and phrases to slow yourself down and to read the questions more carefully.

_____ Analyze the relationships expressed in the questions.

_____ Look for word clues.

_____ Visualize or recall notes, figures, and summaries to help answer questions.

_____ Try to think of an answer to a question before reading the choices given in the question.

_____ Read all choices in a multiple-choice question.

_____ Try to find answers to unknown questions on other parts of the test.

_____ Eliminate obviously wrong answers in multiple-choice questions before selecting the best choice.

_____ Review the whole test if enough time is available.

_____ Analyze questions that were wrong on a returned test.

_____ Try to identify why you gave the wrong answers on a marked test.

_____ Keep old tests for review during preparation for the final examination.

14

Test-Taking—Analyzing Results

Objectives

1. To learn an overall approach to taking tests.
2. To learn specific test-taking skills.
3. To learn how to analyze returned tests and to realize that this analysis can help you prepare for future tests.

Taking Tests

When taking a test, you should adopt a standard and organized test-taking procedure. This procedure involves proper preparation, a confident approach, good test-taking skills, and an analysis of returned tests (figure 14.1). The approach

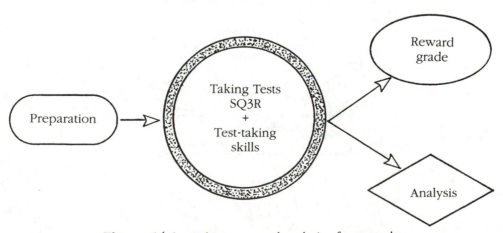

Figure 14.1 Taking tests and analysis of test results.

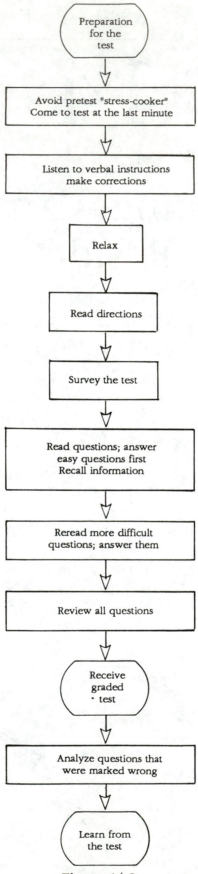

Figure 14.2

to answering essay and math problem-solving questions was discussed in chapter 12. A flow chart in figure 14.2 represents an overall procedure successful students follow when they take a short-answer test.

All tests make students a bit anxious; this is a natural reaction. If you study well, then you should have confidence that you will do well on the test. There is no need to feel overly anxious (easy to say but sometimes hard to do). If you feel very anxious about a test, then you should try to arrange to relax before the test by reading the end of a good novel or an interesting short story. Get your mind off the academic challenge (appendix 2 p. 130).

Plan to arrive at the classroom at the last minute. Avoid getting suckered into a "Did you study this, did you study that?" pressure cooker stress session outside or inside the classroom. Remember, you are in charge, you have taken control of your studying and learning.

Accept the test, but avoid the temptation to rush into it. Listen attentively to all instructions and corrections given by your instructor. Make any changes then and there.

A simple relaxation exercise might be worth doing if you are nervous. Breathe in, hold, then exhale on the count of ten. Repeat this, but then envision a peaceful and relaxing scene . . . a back rub, a frog on a lily pad, a quiet day on the river, lake, or ocean. Now you are ready.

SQ3R the test

S—Survey.   Begin the test by surveying the entire test. Take note of the following:

- directions and time allowed for test;
- number and types of questions;
- sequence of test . . . (Does the test follow the lecture sequence, or are the topics mixed up?);
- value of each part of the test.

Q—Questions.   The questions are generated for you, but as you begin to read the instructor's questions, realize that you will want to ask yourself other questions that will help lead you to the correct answer. *Don't make the questions more difficult than they are.*

R—Read.   First read and follow all the directions. Then read each question slowly and carefully.

- Use your pencil and guide yourself through the question.
- Reread the question and underline the keywords.
- Read and answer the easy questions first. Go back to the more difficult questions later for still another reading and analysis.

R—Recall.   After reading each question, recall information from your study about that question. Think of the answer and record it. If a question is difficult and you can't recall the information or can't think of the answer, then don't dwell on it. Mark the question and return to it later. Your mind might subconsciously

QUESTIONS

- How should the answers that ask for definitions or short explanations be phrased?

- What is an example?

ADDED CONTENT

When answering questions in science, you must:

1. Use the vocabulary of science.

2. Compose your answer using the vocabulary of the question.

3. Write complete sentences.

Examples:

Define osmosis.

A good answer is

Osmosis is the diffusion or movement of water from a region of high concentration of water to a region of low concentration of water through a semipermeable membrane or plasma membrane.

A wrong answer is

It is the movement of something through a membrane.

Note that the good answer repeats the word *osmosis,* and uses the words *diffusion, water molecules, semipermeable,* and *plasma membrane.* This answer demonstrates the comprehension of scientific terminology and the concept of osmosis. The wrong answer lacks much of this information.

search for the answer to it. You might find clues to the answer in other questions you read and recall. If you have studied, the questions should be challenging but answerable.

Return to the more difficult questions. Take time to recall or possibly recite information about the topic of the question. Try to eliminate choices that are obviously wrong. Reread the question once again and concentrate on the key-words of the question and the possible choices. Which choice "fits"? Which choice makes a *true* statement when combined with the question? That will be your answer. If nothing rings a bell, then make an educated guess if there is no penalty for guessing on the test.

R—Review.   After answering all the questions, check the time. If you have the time to relax a minute then do so. Think about what you will do after the test . . . what you will do during the coming weekend. Return to the test and review all questions if there is time. Don't change an answer without due cause. Read each and every choice to make sure they read as *true statements.*

If you have the time and the stamina, there is another review you can do when finished with the test. Think of the questions you generated in your notebook. How do they compare with the questions generated by the in-structor? Start to learn how your instructor composes a test, and endeavor to generate questions similar to the ones that appear on the test. Try to keep in mind the more difficult questions. What kind are they? Can you visu-alize the instructor lecturing on these topics?

Test-Taking Skills and Hints

- Survey the test right away.
- Answer the easy questions first. This builds confidence.
- Take the questions for what they are; don't read hidden meanings into the questions.
- Skip difficult questions, but mark which ones you skip so you don't have to search for them. Don't forget to skip the question on the answer sheet.
- Look for clues or answers to a skipped question in the statements contained in other parts of the test.
- Work at a slow, steady pace.
- Use your pencil to guide your reading; read "aloud" if it helps.

Q U E S T I O N S

- What do essay questions ask?

- How should you analyze figure 14.2?

A D D E D   C O N T E N T

There are specific terms that give questions direction. You should know the meaning of these terms. The *information table* (figure 14.2) outlines the vocabulary of essay tests.

The table below outlines the various categories of test terms, tells you the type of answer that is needed, and then gives examples of specific terms that fit into each category.

General Category	Answer Needed	Examples of Specific Terms
Identification	Present the bare facts: a date, a name, a phrase; in short, provide a concise answer.	cite, define, enumerate, give, identify, indicate, list, mention, name, state
Description	Tell about a specific topic with a certain amount of detail.	describe, discuss, review, summarize, diagram, illustrate, sketch, develop, outline, trace
Relation	Describe the similarities, differences, or associations between two or more subjects.	analyze, compare, contrast, differentiate, distinguish, relate
Demonstration	Show (not state) why something is true or false. Put forth logical evidence or arguments to support a specific statement.	demonstrate, explain why, justify, prove, show, support
Evaluation	Give your opinion or judgment on a subject plus justify and support it. Also, if your opinion can be challenged, be sure to present both sides.	assess, comment, criticize, evaluate, interpret, propose

Figure 14.3   Note this table is about the vocabulary of essay test taking. Compare these general categories to the terms in the matching exercise on page 28 of chapter 6. (Source: The definitions of answers needed and examples of specific terms are adapted from Jason Millman and Walter Pauk, *How to Take Tests* (New York: McGraw-Hill, 1969), pp. 152–57. Used with permission. The form of figure 14.3 itself is taken from "The Vocabulary of Test-Taking." Reprinted by permission from the 1975/76 issue of *Nutshell*. Copyright © 1975 by 13–30 Corporation.)

- Underline keywords to help yourself concentrate on the question and the choices.
- Don't let the pace of others influence your test-taking procedure; use all the time given to you if you need it.
- Watch for word clues. Statements that express absolute quantity (*all, always, exactly, none, never*) are usually false. True statements use *some, often, many,* or *on the average.* If the question asks about pressure, then don't select an answer that involves volume or temperature.
- On *true-false* tests, if *any* part of the statement is false, the whole statement is false. Don't be tempted to answer true because it sounds good; all words and phrases must "fit" for the statement to be true.

- *Multiple-choice* questions:
- After reading the question, try to think of the answer.
- Read all the choices to find the correct answer.
- Note whether the choices include a combination of answers. Be doubly careful to read all choices. The instructor might be asking you to combine information from different lectures in this type of question.
- It is helpful to cross out choices that are obviously wrong, leaving the possible choice to think about.
- Correct answers will read as a true statement. (All other choices will make the question into a false statement.)
- If you don't know the answer, guess at it after eliminating the obviously wrong answers.

- *Matching* questions:
- Read each column.
- Reread each choice in the left-hand column and recall information about the statement or word.
- Think of possible answers and look for the answer in the right-hand column.
- Cross out used choices if the directions say that possible answers can be used only once.
- Remember to do the easy ones first, then the more difficult ones last.

QUESTIONS

- How can you use your textbook and laboratory manual to help you study for tests?

ADDED CONTENT

1. The glossary helps you define keywords.

2. Headings and subheadings will help identify important information in the text and manual.

3. Headings and subheadings can be turned into sample test questions.

4. The bold-printed or italicized words are important words to add to your vocabulary.

5. The chapter summaries outline the chapters' content.

6. The figures help you envision the discussion in the text and the experimentation in the lab. If you know how to explain the figures, then you will know much of the content.

7. The sample solutions help you form the basis to solve problems at the end of the chapter.

8. Treat end-of-chapter or end-of-experiment questions as test questions. They will help you direct your study. Concentrate on the questions that relate to the topics discussed in lecture or the experiments performed in laboratory.

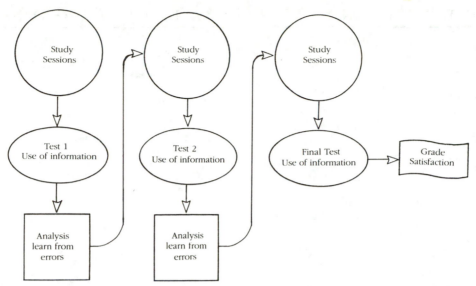

Figure 14.4   Sequence of study, testing and analysis of results which lead to a grade and to some degree of personal satisfaction.

Hand in the test.   Perhaps you'll be able to keep the test questions. If not, then realize that when you get your corrected answer back you'll have to make an appointment with the instructor to compare the corrected test to the question. A review of your mistakes is an important step in learning how to study and how to take future tests. You want to learn from your mistakes, not reinforce the process of making mistakes.

Analysis of Test Results

When the test is returned, you may feel satisfied with the results. If you are not satisfied, then you must realize you have to do something to increase your level of satisfaction. If you did not prepare for the test, then you should not expect much satisfaction. Regardless of the results, whether you got one or twenty answers wrong, you should analyze those questions you got wrong. You should find out why you answered the questions incorrectly.

Why is the Answer Wrong?

Look at each question you have gotten wrong on your test. Try to determine why the answer is wrong. What was the most frequent cause of errors? Try to correct those weaknesses in the future. The following are some reasons you might have answered the questions wrong:

1. You misread the question or you read the question carelessly.
2. You did not read all of the choices.
3. You did not know or identify the keywords in the question.

Q U E S T I O N S   A D D E D   C O N T E N T

4. You were unfamiliar with the information in the question and all of the choices.
5. You could not use or apply the information you knew (or memorized) to a question.
6. You added your own meaning to the question.
7. You recorded the wrong answer on the answer sheet.

Comparing Mistakes to Your Notebook

Compare the questions you got wrong to the content of your notebook and weekly summaries. Can you find the answers in the notebook and summaries? Try to judge whether your notes are complete. If information is missing from your notes, then you must ask whether the instructor covered the material or whether your notes are complete or incomplete. Plan to improve your note-taking skills if that seems to be the problem. If you think the instructor did not cover the material, then meet with the instructor to discuss why she thinks you should have been able to answer that question. Did you study your notes enough? Did you forget what was in your notes?

Comparing Mistakes to the Textbook

Try to determine where the content of the questions you got wrong can be found in the text. Had you read that part of the text? Was the information part of a reading assignment? Did you consider the information in the text too unimportant to study? Was the information in the question part of the headings or subheadings of the textbook?

Comparing Test Questions to Your Questions

The last comparison you should do is between the instructor's questions and your questions. How did the questions you recorded in your notebook compare to the instructor's test questions? Did the types of questions you asked yourself help or hinder you in studying for the test? What kinds of information did the instructor ask questions about: facts, concepts, or application of information?

Results of Analysis

After analysis of your marked test, you should know why you answered various questions incorrectly. You should also have identified whether your notes are complete or not, whether you are using the textbook efficiently, and whether or not you have studied effectively. Resolve to correct the deficiencies as much as you can or want to. Take better notes, review the text, prepare your own questions, read, review, and recall more frequently, make up more effective summaries, information charts, and maps, practice problem-solving. In other words, study more efficiently and effectively.

If you can't determine the source of your mistakes on the test, bring the test and the questions to your instructor or science tutor. Ask for their help and guidance.

Review

1. Prepare for the test; this should begin two or three days before the test.
2. Try to relax before the test.
3. Come to the test on time, but not too early.
4. Apply the SQ3R reading technique to your test-taking.
5. Use good test-taking skills; read carefully and slowly, read everything, visualize notes, figures, and summaries to help you answer the questions.
6. Review the test before handing it in.
7. Review the test mentally after handing it in. Think of how it compared to your notes and summaries.
8. Analyze the test results. Determine why you got things wrong; compare the test to the way you prepared for the test.
9. Ask your instructor for help if you need to.

A P P E N D I C E S

Contents

1
Answers to selected science questions

2
Anxiety and your study of science

3
Mnemonics

APPENDIX 1

Answers to Selected Science Questions

Many questions have been asked in the *questions column* of this book. You should probably be able to answer most of these questions after reading the content or thinking about how you do things. There are a few places where the text asked you specific things about science. The answers to these questions are given here to allow you to check your answers.

These are the symbols represented in figure 10.1.

Answers to weather-map questions, figure 10.3, page 74

Note how the vocabulary of the question is incorporated in the answer. The answers are also complete sentences.

1. Boston's weather is rainy and the temperature is in the 40s.

2. The temperature in Washington, Chicago, and Seattle is in the 60s.

☼ – sun

N, E, S, W – North, East, South, West

∴ – Therefore

ATP – adenosine triphosphate

≺ – vectors

u – micron, micrometer

O_2 – oxygen gas, 2 atoms of oxygen

∿ – warm front

F = ma – Force equals mass times acceleration

⊕ ⊖ – hydrogen atom
⊕ - positive charge proton, nucleus
⊖ - negative charge electron
⌒ - orbit

~ – approximately

> – less than

H_2O – water molecule 2 Hydrogen atoms 1 Oxygen atom

[ ] – concentration

S‰ – salinity

°F – degrees Fahrenheit

⊙ – a cell, nucleus and membrane ∴ = organelles

°C – degrees Celsius

kcal – kilocalorie

glucose – an organic compound, a simple sugar

[P] – Producer biomass

[C] – Consumer biomass

[D] – Decomposer biomass

→ – flow of energy or nutrients

$C_6H_{12}O_6$ – molecule of glucose 6 atoms hydrogen 12 atoms hydrogen 6 atoms oxygen

ml – milliliter

m – mass, meter

Cu – copper, copper atom cumulus cloud

⬡ – molecular structure of glucose – short form

$HCl \rightleftharpoons H^+ + Cl^-$
Hydrochloric acid molecule dissociating into H^+ (hydrogen ion) and Cl^- (chloride ion)
→ dissociation tends to form ions more strongly than re-association to form molecule.

3. A cold front is east of the east coast. A warm front is just east of Boston.

4. The dot-and-dash lines connect areas of the country having the same ranges of temperature. For example, there are bands of 80s, 70s, 60s, and 50s running roughly west to east.

5. Long Island, New York, was in the 50s.

6. Snow fell in central Montana just northwest of Billings.

Answer to the representation of the genetic symbols in chapter 10 figure 10.2 page 73.
The symbols represent a mating between two genetic types (genotypes) of organisms (rr and Rr) and display the genetic types of the offspring.

The lowercase "r" represents a recessive allele (gene); the "R" allele is dominant to the "r." Genotypes of the parents are represented as "rr" and "Rr." The former is said to be homozygous recessive, the latter heterozygous. The homozygous recessive genotype can form only one type of gamete (reproductive cell), with only the recessive allele in the gamete. The heterozygous genotype can form two types of gametes—one with a dominate allele in it, the other with a recessive allele.

These gametes are arranged in what is called the Punnett square, in which the gamete types are hypothetically mated to form new gene arrangements when fertilization occurs. The new generation is represented in the Punnett square as "Rr" and "rr" in a 1-to-1 ratio. Thus, heterozygous and homozygous recessive parents can produce only offspring that are heterozygous and homozygous recessive. Each genotype has a 50 percent chance of appearing in the offspring.

Answers to the questions about the figure 11.7 on page 88.

2. A balanced chemical equation is a chemical equation in which the number of atoms on the left is equal to the number of atoms on the right.

3. An arrow separates an equation into a left-hand and a right-hand side.

4. The numbers of atoms for each element are the same on the left-hand and right-hand sides of a chemical reaction.

5. H represents one atom of the element hydrogen.
    H_2 represents one molecule of hydrogen gas, which is made up of two atoms of hydrogen.
    O represents one atom of the element oxygen
    O_2 represents one molecule of oxygen gas, which is made up of two atoms of oxygen.
    H_2O represents one molecule of water, which is made up of two atoms of hydrogen and one atom of oxygen.

6. H_2 is represented by two interconnected shaded spheres; each sphere is a hydrogen atom.

7. O_2 is represented by two larger unshaded, interconnected spheres; each sphere is an oxygen atom.

The water molecule is represented by one large sphere (oxygen) to which two smaller spheres (hydrogen) are bonded.

The atoms overlap to indicate that the atoms are chemically bonded to each other.

8. The (+) means that the two hydrogen molecules and one oxygen molecule react together. The arrow represents the fact that these reactants are changed in a chemical reaction to the product that is located to the right of the arrow.

9. The diagram "a" attempts to help you visualize the atoms in molecules that are reacting in a balanced chemical reaction to form a product. The molecules that are reacting to form two molecules of water are symbolized in "b." In "c" the balanced reaction is represented in its briefest form. The coefficient of 2 is placed in front of the hydrogen gas and water molecules to represent the numbers of those molecules involved in the reaction.

10. The 2 in front of the H_2 and H_2O indicates that two of each of these molecules are involved in the reaction.

12. The hydrogen atoms separate from each other and the oxygen atoms separate from each other. As the atoms react, two hydrogen atoms bond to one oxygen atom. This happens twice in the balance reaction.

13. The natural shape of the molecule has been found to be this way. That is just the way it is. Of course it has to do with the interrelationships of the subatomic parts of the atoms involved.

14. Two molecules of hydrogen gas will react with one molecule of oxygen gas to form two molecules of water. Each molecule of water has two atoms of hydrogen and one atom of oxygen.

15. Keywords are *balanced chemical equation, hydrogen gas, oxygen gas, water, atom, molecule, chemical reaction.*

Answers to the sample test questions in chapter 14, starting on page 110.
Oceanography: 1. F, 2. c, 3. d
Meteorology: 4. T, 5. b, 6. d,
Geology: 7. F, 8. c, 9. a
Biology: 10. T, 11. a, 12. e
Chemistry: 13. T, 14. a. 0.1M, b. 0.1M, c. 0.2M, d. 0.3M

Appendix 2

Anxiety and Your Study of Science

Some people become so anxious about studying science or taking a test that "their minds go blank" or "they can't think straight." Often these same people feel the science course (or math course) required by the curriculum will make or break their college careers.

If you are extremely anxious about the science course you are or will be studying, then there are at least two things you can do. The first is to visit a counselor on campus to help you deal with your real but inhibiting state of mind. Frequently colleges have counselors who have specialized in anxiety and stress management. Visiting them could help.

The second thing to do is to concentrate on learning the skills and taking the time necessary to study science. Do not dwell on the reasons you can't study, can't learn, or can't take tests. This negative "whirlpool" can consume excessive time and energy.

Practice and use the methods of study recommended in this book. Begin to prepare for the tests on the first day of class. Realize that if you do this conscientiously, you should feel confident about your ability to pass the course. If you learn from the first test, as suggested in chapter 16, then you should do better with each passing week and each succeeding test. Seek out or form a supportive study group.

From time to time, you might feel you need a "crutch" to help you reduce your level of anxiety. You can do that with practice. In 1966, Joseph Wolpe devised the Subjective Unit of Disturbance Scale (SUDS) to help people gauge and reduce their anxiety.

Subjective Unit of Disturbance Scale (SUDS)

Worst anxiety experienced or imagined

Absolute calm peace

Think of two experiences or states of mind that yield two extremely different levels of anxiety. The first is an anxiety-free state, a scene that yields a peaceful state of mind. Register this as "0" on the SUDS scale. Next, think of the most anxious experience or scene that you can imagine. This "worst of worst" experiences should be "100" on the SUDS. From this point on, you can now evaluate your anxiety level on your SUDS "meter" during class, while studying for tests, or just before tests. Once you do this, you can take measures to reduce your level of anxiety. Learn a relaxation technique, and apply it to reduce your SUDS meter reading. Talk to yourself. Motivate yourself to act positively. It might help if you practice this. This might help reduce your science anxiety enough to give you the confidence you need.

There is a risk involved in studying science or any other subject. A positive self-image and feeling of self-worth will make you feel confident that the risks are really relatively small. This is true only if you learn how to learn, study, and be active in learning.

You must remember that low anxiety levels and poor preparation will not enable you to do well in a science course. Quality study time and quality preparation are the cornerstone to your success in a science course. Applying good study skills for a long enough time should enable you to pass the course and to reduce your levels of anxiety.

Appendix 3

Mnemonics

Mnemonics are short verbal devices that help you recall a series of facts. The verbal device is a code to the series of facts you must learn. You can make up your own mnemonics or learn to establish ones. Mnemonics can be silly sentences, jingles, or acronyms. An acronym is a word formed from the initial letters or letters of each of the successive parts or major parts of a compound term or series of facts.

Silly sentence mnemonics
Geology

"*C*amels *O*ften *S*it *D*own *C*arefully. *P*erhaps *T*heir *J*oints *C*reak. *P*ersistent *E*arly *O*iling *M*ight *P*revent *P*ermanent *R*heumatism.

The first letter of each word refers to the geological time periods:

Cambrain, Ordovician, Silurian, Devonian, Carboniferous, Permian, Triassic, Jurassic, Cretaceous, Paleocene, Eocene, Oligocene, Miocene, Pliocene, Pleistocene, Recent.

Anatomy

"*L*azy *F*rench *T*arts *S*it *N*aked *I*n *A*nticipation" encodes some of the nerves passing through the skull.

Lachrymal, frontal, trochlear, superior, nasal, inferior, and abducent.

Physics

"*V*irgins *A*re *R*are."

In the study of electricity, Ohm's law states that *volts* are equal to *amperes* times *resistance* (v = ar).

Some people find mnemonics helpful; others find they just add to the things one must remember. Students working in study groups seem to use and share the humorous mnemonics more frequently. Maybe you have to hear them to believe them. You should definitely consider using mnemonics as a tool to help you learn material and prepare for tests.

Quick study skills lists

What is important to study?

1. Anything mentioned in lab and lecture.
2. More time on topic = more important.
3. If in lecture, lab, and text, then very important.
4. All scientific words and symbols used by instructor.
5. Content of figures used by instructor.
6. Problems solved in class and on assignments.

Surveying

1. Look over headings, subheadings, and block-print words.
2. Read the introduction and chapter summary.
3. Scan the figures; get a general feel for them.
4. Try to predict what is important.

Reviewing

1. Review notes within twenty-four hours.
2. Highlight or underline keywords, symbols, equations.
3. Make corrections, fill in gaps.
4. Write a summary of the lecture.
5. Study material in text that relates to lecture notes.
6. Outline text information in the added content column of notes.
7. Analyze all figures.
8. Compare definition in notes and text.
9. Make up flash cards.
10. Compose information maps, tables, charts.
11. Study with classmates.

Sources of questions

1. Questions the instructor asks in class.
2. Questions you think of while listening to instructor or studying.
3. Conversion of headings, subheadings, and bold-print words into questions.
4. End-of-chapter questions.
5. Study guide.
6. Other students in study group.

Analysis of figures.

1. Survey the figure. What is it about?
2. Read the caption to answer what it is about. Underline keywords and review their meanings.
3. Clarify meaning of words or concepts.
4. Relate the caption to the figure content.
5. Identify and interpret all parts of the figure.
6. Ask and answer questions about the figure.
7. Relate the content of the figure to the content of the lecture.

BIBLIOGRAPHY

Ambron, J. "Writing to Improve Learning in Biology." *J.C.S.T.,* February 1987, 263–266.

Annis, L. F. *Study Techniques.* Dubuque, Iowa: Wm. C. Brown, 1983.

Bogue, C. *Studying the Content Areas.* Clearwater FL: H & H Publishing, 1988.

Bosworth, S., and M. A. Brisk. *Learning Skills for the Science Student.* Clearwater, Florida: H & H Publishing, 1986.

Carver, J. B. "Ideas of Practice: Plan-Making: Taking Effective Control of Study Habits." *Journal of Developmental Education,* November 1988, 5(2) 26–29.

Cooke, L. M. "Design for Excellence. How to Study Smartly." National Action Council for Minorities in Engineering, Inc. New York.

Crafts, K., and Hauther, B. *Surviving the Undergraduate Jungle.* New York: Grove Press, 1976.

Davis, A., and Clark, E. G. *T-Notes and Other Study Skills.* Metamora, Illinois: Davis & Clark Publishing, 1985.

Elliott, H. C. *The Effective Student.* New York: Harper & Row, 1966.

Farrar, R. T. "College 101." *Petersons Guides.* Princeton, New Jersey, 1984.

Foo, S. *Noteworthy Success.* Cornell Countryman, Vol. 36, pp. 11, October 1988.

Fry, R. W. *How to Study.* Hawthorne, New Jersey: The Career Press, 1989.

Geoffrion, S. *Get Smart Fast: A Handbook for Academic Success.* Saratoga, California: R. & E. Publishers, 1986.

Haburton, E. "Study Skills Packet." Valencia, Valencia Community College, California, 1978.

Heiman, M., and Slomianko, J. *Methods of Inquiry.* Cambridge, Massachusetts: Learning Associates, 1986.

Kay, R. S., and Terry R. A., Ed. *How to Stay in College.* Washington, D.C.: University Press of America, Inc., 1978.

Kesselman-Turkel, J., and Peterson, F. *Study Smarts. How to Learn More in Less Time.* Chicago: Contemporary Books, Inc., 1981.

Kesselman-Turkel, J., and Peterson, F. *Note-taking Made Easy.* Chicago: Contemporary Books, Inc., 1982.

Knowles, M. *Self-directed Learning.* New York: Association Press, 1975.

Langan, J. *Ten Steps to Improving Reading Skills.* Cherry Hill, New Jersey: Townsend Press, 1988.

Maiorana, V. P. *How to Learn and Study in College.* Englewood Cliffs, New Jersey: Prentice-Hall, Inc., 1980.

Mallow, Jeffery. *Science Anxiety.* Clearwater, Florida: H & H Publishing, 1977.

Maxwell, Martha. *Improving Student Learning Skills.* San Francisco: Jossey-Bass Publishers, 1981.

Miles, C., and Rauton, J. *Thinking Tools.* Clearwater, Florida: H & H Publishing, 1985.

Ohm, H. *Note Taking and Report Writing.* Palo Alto, California: California Education Plan Inc., 1989.

Strauss, M. J., and Fulwiler, T. "Interactive Writing and Learning Chemistry." *J.C.S.T.,* February 1987, 256–262.

Tonjes, M. J., and Zintz, M. V. *Teaching Reading, Thinking, Study Skills in Content Classrooms.* Dubuque, Iowa: Wm. C. Brown, 1981.

Trillin, A. S. and Associates. *Teaching Basic Skills in College.* San Francisco: Jossey-Bass, 1980.